Optimization of Pharmaceutical R&D Programs and Portfolios

Zoran Antonijevic
Editor

Optimization of Pharmaceutical R&D Programs and Portfolios

Design and Investment Strategy

Springer

Editor
Zoran Antonijevic
Cytel Inc.
Cambridge, MA, USA

ISBN 978-3-319-09074-0 ISBN 978-3-319-09075-7 (eBook)
DOI 10.1007/978-3-319-09075-7
Springer Cham Heidelberg New York Dordrecht London

Library of Congress Control Number: 2014949773

Printed on acid-free paper

Springer is part of Springer Science+Business Media (www.springer.com)

Contents

Contributors

Suresh Ankolekar Maastricht School of Management, Maastricht, The Netherlands

Zoran Antonijevic Cytel Inc., Cambridge, MA, USA

Robert A. Beckman, M.D. Daiichi Sankyo, Blue Bell, PA, USA

Sarah T. Bobulsky Ernst and Young, LLC, New York, NY, USA

Jim Bolognese Cytel Inc., Cambridge, MA, USA

Carl-Fredrik Burman AstraZeneca R&D, Molndal, Sweden

Cong Chen, Ph.D. Merck Research Laboratories, PA, USA

Christy Chuang-Stein Pfizer, Kalamazoo, MI, USA

Frank S. David, M.D., Ph.D. Pharmagellan, LLC, Milton, MA, USA

Raymond A. Huml, M.S., D.V.M., R.A.C. Quintiles Transnational Corp., Durham, NC, USA

Blair Ireland Novartis, East Hanover, NJ, USA

Chris Jennison Department of Mathematical Sciences, University of Bath, Bath, UK

C. Kwon Kim, Ph.D. Kromite, Vancouver, BC, Canada

Martin Kimber Tessella, Abingdon, Oxfordshire, UK

Jack Kloeber Jr., Ph.D. Kromite, Feasterville, PA, USA

Olga Marchenko Quintiles Innovation, Morrisville, NC, USA

Judith Ng-Cashin, M.D. Quintiles Transnational Corp., Durham, NC, USA

Richard Nixon Novartis Pharma AG, Basel, Switzerland

Nitin R. Patel, Ph.D. Cytel Inc., Cambridge, MA, USA

Charles Persinger Eli Lilly and Company, Indianapolis, IN, USA

José Pinheiro Janssen R&D, Raritan, NJ, USA

Frederic (Rick) Sax, M.D. Quintiles Transnational Corp., Durham, NC, USA

Kraig F. Schulz Ernst and Young, LLC, Bridgeport, CT, USA

Alex Stojanovic, Ph.D. Phosphagenics, Lawrence Township, NJ, USA

Linda Z. Sun, Ph.D. Merck Research Laboratories, PA, USA

Part I
Introduction

Chapter 1
Need for Optimal Design of Pharmaceutical Programs and Portfolios in Modern Medical Product Development

Zoran Antonijevic

Portfolio optimization is the process of managing its components such that the output is maximized based on a selected criterion. In pharmaceutical industry portfolio components are individual development programs. The selected criterion is usually a financial measurement, such as expected net present value (ENPV). There are numerous constrains in optimizing pharmaceutical portfolios; either at the portfolio level (budgets), or at individual components level (regulatory, payors).

Very little has been published on optimization of pharmaceutical portfolios. Moreover, most of published literature is coming from the commercial side, where probability of technical success (PoS) is treated as fixed, and not as a consequence of development strategy or design. In this book there is a strong focus on impact of study design on PoS, and ultimately on the value of portfolio. Design options that are discussed in different chapters are dose-selection strategies, adaptive design, enrichment, and selection of sample size. Some development strategies that are discussed are indication sequencing, optimal number of programs, and optimal decision criteria.

This book includes chapters written by authors with very broad backgrounds including financial, clinical, statistical, decision sciences, commercial, and regulatory. Many authors have long held executive positions and have been involved with decision making at a product or at a portfolio level. As such, it is expected that this book will attract a very broad audience, including decision makers in pharmaceutical R & D, commercial, and financial departments. The intended audience also includes portfolio planners and managers, statisticians, decision scientists, and clinicians.

Early chapters describe approaches to portfolio optimization from big Pharma and Venture Capital standpoints. They have stronger focus on finances and processes. Later chapters present selected statistical and decision analysis methods

Z. Antonijevic (✉)
Cytel Inc., 675 Massachusetts Ave, Cambridge, MA 02139, USA
e-mail: zoran.antonijevic@cytel.com

Z. Antonijevic (ed.), *Optimization of Pharmaceutical R&D Programs and Portfolios: Design and Investment Strategy*, DOI 10.1007/978-3-319-09075-7_1

for optimizing drug development programs and portfolios. Some methodological chapters are technical; however, with a few exceptions they require a relatively basic knowledge of statistics by a reader.

Pharmaceutical Industry Challenges

The Pharmaceutical industry has come to an era with increased pressure on sponsors to improve the cost-effectiveness, productivity and quality of their product development. Recently they have been facing increasing costs and diminishing returns. There are many reasons for this, and we need to understand them before we start proposing solutions.

First, the era of blockbuster drugs is coming to an end. With the blockbuster model companies would target diseases that affect large populations; like heart disease, pain, or depression just to name a few. Often, companies would develop drugs from the same class with already approved drugs, for example statins, cyclooxygenase 2 (COX-2) inhibitors, or selective serotonin reuptake inhibitors (SSRIs). Alternatively they would target a drug that is just marginally better than current treatment. The market for above mentioned diseases was so large that even sharing it would result in making billons annually. The development path would not have to be very innovative, and development risks were relatively marginal. Additionally, companies were able to slightly change their drugs, extending their patent time by years and selling them as new treatments. It is easy to see why this model was very successful for a long period time. There are many reasons why the above described model is no longer prevalent, the Affordable Care Act being the most significant one. First, it is now required from new drugs to significantly outperform already available products so that they can be reimbursed by the insurance. Furthermore, it is now much easier to get the approval for generic versions of drugs, and they are becoming fierce competitors.

The emergence of safety issues with some blockbuster drugs, like Vioxx and Avandia, has resulted in more stringent safety requirements for approval of drugs. The best example of this is the FDA Guidance for evaluating cardiovascular risk for new therapies for treatment of Type II Diabetes. This guidance essentially requires the addition of a cardiovascular outcomes study with thousands of patients in order to achieve its requirements. Needless to say, the impacts of such a guidance are increased development costs and risks, and delayed approvals.

Finally, the primary focus of drug developers has for a long time been on the cost and the speed of development. The most important parameter, the Probability of Success (PoS) has largely been overlooked, and drugs were developed as if the success was imminent. The PoS is the most important factor for a number of reasons. First, without a successful submission there will be no revenues. With no revenues failing drugs are adding to the average cost of development, and the later they fail the more the cost. Interestingly, not only do companies assume that their drug will succeed, but they also develop strategies as if the competitor will succeed with certainty.

This resulted in drug development becoming a chase of who will get to the end faster, without paying much attention to how to improve PoS and benefit–risk profile through study design and sound development strategy. An obvious, but rarely utilized alternative to chase is differentiation. Under this alternative sponsors' primary focus would be on studying an optimal dose, regiment, or a subpopulation that would experience greatest benefits from treatment. When pharmaceutical companies evaluate PoS, it is typically evaluated in a subjective process by a committee of experts making arbitrary guesses, with little use of the data generated in the program to that point. Another variant is input of these guesses into programs which compute the probability associated with a tree, but without requiring a basis for the inputs.

To sum up: pharmaceutical companies are facing many new challenges. From the regulatory standpoint there are new safety and efficacy requirements that will result in additional clinical trials, increase in cost of drug development, delay of submissions, and increased difficulties to get regulatory approval. The new challenge for any newly approved drug is to get reimbursed. Therefore, differentiation is another key element. Instead of targeting entire populations affected by a very prevalent disease, companies will have to plan and invest a lot more in identifying target populations that can benefit the most from their treatment. In so doing they will have to select from a portfolio of putative predictive biomarkers and classifiers, and this is also best done in a quantitative data driven fashion designed to maximize expected utility [1]. Finally, new drugs are facing more competition, and less flexibility for extending patent times.

Clearly the current process of decision making, which has for a long time been siloed within individual departments, and based on executives' "gut feeling" is no longer sustainable. There has to be much more focus on quantitative decision making based on measurable parameters. The ultimate goal for drug developers has to be to maximize the expected value of their products, and portfolios.

Lack of Quantitative Decision Making in the Pharmaceutical Industry

The pharmaceutical industry is far behind other major industries when it comes to use of quantitative methods to support decision making. Examples of this are lack of utilization of decisions analysis, inadequate use of statistical resources, and lack of utilization of Modeling and Simulation to support decision making. These are discussed in more detail later in this section.

One possible explanation for a lack of scientific approach in decision making is that historically, and particularly during the blockbuster era, getting marketing approvals was less challenging while revenues for approved drugs were huge. As a result, Pharmaceutical companies' profits were large, and executives were not compelled to change much in their decision-making style. Another commonly mentioned explanation is that Pharmaceutical industry leaders are not quantitative people, and as such are reluctant to base decisions on quantitative methods or simulation outputs.

Finally, the intent of quantitative methods, and simulations in particular is to provide solutions based on the totality of evidence. The pharmaceutical industry is very compartmentalized and different groups tend to downplay the value of others. Throughout this book we discuss the value of integrated quantitative approaches.

Decision Analysis Not Sufficiently Utilized

The use of decision analysis would allow pharmaceutical companies to maximize the expected utility and value of their portfolio in a systematic, data-driven fashion. Currently, most companies prioritize portfolios based on arbitrary criteria and judgments, and typically draw a funding line. In some cases the funding is stratified such that some projects are designated high priority and receive more funds to accelerate development. PoS for these high priority assets is often inflated based on gut feelings, so that the perceived value determines the PoS rather than the other way around. Instead, pharmaceutical companies need to make use of available data to estimate key parameters, and perform quantitative trade-off analysis between development options and quantitative trade-off analysis across portfolios, with the aim of maximizing expected value or utility.

Inadequate Use of Statistical Resource

The pharmaceutical industry employs a very large number of statisticians with advanced degrees, yet their input is restricted to routine tasks that can be done by people with much less quantitative background and sophistication. For example, most common tasks for statisticians in the pharmaceutical industry are data manipulation and production of routine tables. In other industries similar tasks are done by people with much less training and education.

Consequentially, statisticians are insufficiently utilized as the strategic resource. Yet they are uniquely equipped with skills that are critical for making strategic decisions, such as: understanding and ability to quantify uncertainty, ability to quantify risk and provide solutions to mitigate risk, or provide optimal solutions. They are also equipped with skills and knowledge to assess how different development and design options would impact the likelihood of a product approval.

Lack of Utilization of Modeling and Simulations (M & S)

Pharmaceutical development is a very complex process that involves considerable uncertainty and very long periods until completion of programs. Because of this mathematical solutions to problems are not always possible and different development strategies and decision options can be compared only by utilization of

M & S. Additionally, M & S provide a framework that can facilitate communication between stakeholders. M & S also allow for preclinical and early clinical findings to be incorporated in "prior distributions" for the assumed efficacy and safety profile of the treatment. This prior knowledge can then be combined with the observed data into the posterior information, using the Bayesian paradigm. M & S can be used to support decision making at any stage of the development, and can accomplish the following:

1. Integration of information from multiple areas, from preclinical development through the submission stage. Outputs from earlier stages can be used to define assumptions for simulation parameters. Finally, simulations allow for incorporation of commercial outcomes through the incorporation of utility functions and the optimization of the ENPV, and as such accommodate the integrated development approach.
2. Assessing multiple scenarios such as differing study designs and endpoints, or even more diverse inputs such as cost of study start-up, accrual rates, and per-subject costs.
3. M & S offer an approach to deal with the computational complexity of maximizing ENPV subject to various constraints, and can be used as a tool for drug development optimization.
4. Accounting for uncertainty.
5. Distributional output.

As such, simulations are very well suited for planning and optimization of pharmaceutical products or portfolios. Please see [2–4] for further discussion and/or examples.

Assessing the Value of a Pharmaceutical Product

Components

There are three key components for assessing the value of a pharmaceutical product:

1. Cost

   Factors that affect the overall cost include subject recruitment, investigator and clinician costs, pharmaceutical product, monitoring costs, data analysis and reporting, interaction with regulatory authorities, administrative costs, and many others.
2. Expected revenues

   Recently the most significant factor impacting the revenues is whether the product will get reimbursed or not. Other factors include indication/affected population size, class/asset share, expected treatment duration per patient, remaining patent time, external market dynamics, and compliance.
3. Risk, or inversely, the PoS of a drug development program

   The PoS represents a product of probabilities of progressing from one stage to another over the course of drug development: early discovery, clinical development phases, regulatory approval, product launch, and commercialization.

Expected Net Present Value as a Metric for Valuation

In development of a drug, considerable resources are invested up front with the expectation of recovering costs and accruing revenues during the later commercialization phase if marketing authorization is granted. Net Present Value (NPV) is a financial measurement tool that is widely used to evaluate future returns. It represents the difference between the present value of the future returns from an investment and the amount of investment, as described in standard textbooks on financial management [5]. However, as the realization of returns depends on successful development, NPV needs to be extended to apply to situations involving various forms of risk (e.g., a product not being approved). A straightforward extension is the expected value of NPV (ENPV) which represents the NPV weighted by development risks, and as such incorporates all three components mentioned in the previous section. The use of ENPV has long been recognized as a financial tool in portfolio management and valuation of investments.

There are many advantages of using the ENPV as the outcome of interest for valuation in drug development. One is that the ENPV naturally accommodates optimization. As illustrated in Fig. 1.1a the function of PoS vs. sample size is a monotonically increasing one. The function of ENVP vs. sample size (Fig. 1.1b) would have an inflection point where the value is maximized. We can then think of the corresponding sample size as being an optimal one. But merely maximizing the ENPV without taking into account the uncertainties of inputs does not adequately characterize the level of risk [6].

There are several other considerations for ENPV. For most indications the cost of drug development is a small fraction of what would be the realized revenues. Therefore the PoS and factors impacting the revenues, time of development in particular, would have much larger impact than costs. Cost is, however, a very important factor to consider, given that in the real world budgets are limited. Consider Fig. 1.1b again; it is very possible that improvements in the PoS could drive the inflection point to the far right, where desired sample size would correspond to investments that exceed available resources, or is in the region where the investment would be so large that it would make the sponsor uncomfortable to invest. If a single program takes a large fraction of the fixed, limited resources, there is an opportunity costs in that other studies, which may have been productive cannot be funded. This opportunity cost has been termed "Type III error" in analogy to the Type I and Type II errors associated with false positives and negatives in the frequentist statistical paradigm [1]. Consideration of Type III error is a critical component of portfolio optimization, yet may be neglected in favor of pouring resources into a limited number of opportunities, especially near-term opportunities.

ENPV has also being criticized as a very unreliable measurement. This is driven by uncertainties in all three components of ENPV, but mostly expected revenues. Naturally, uncertainties are larger during earlier stages of development, and several chapters of this book address ways to deal with these uncertainties. We are by no mean suggesting that ENPV should be the only outcome of interest in optimizing

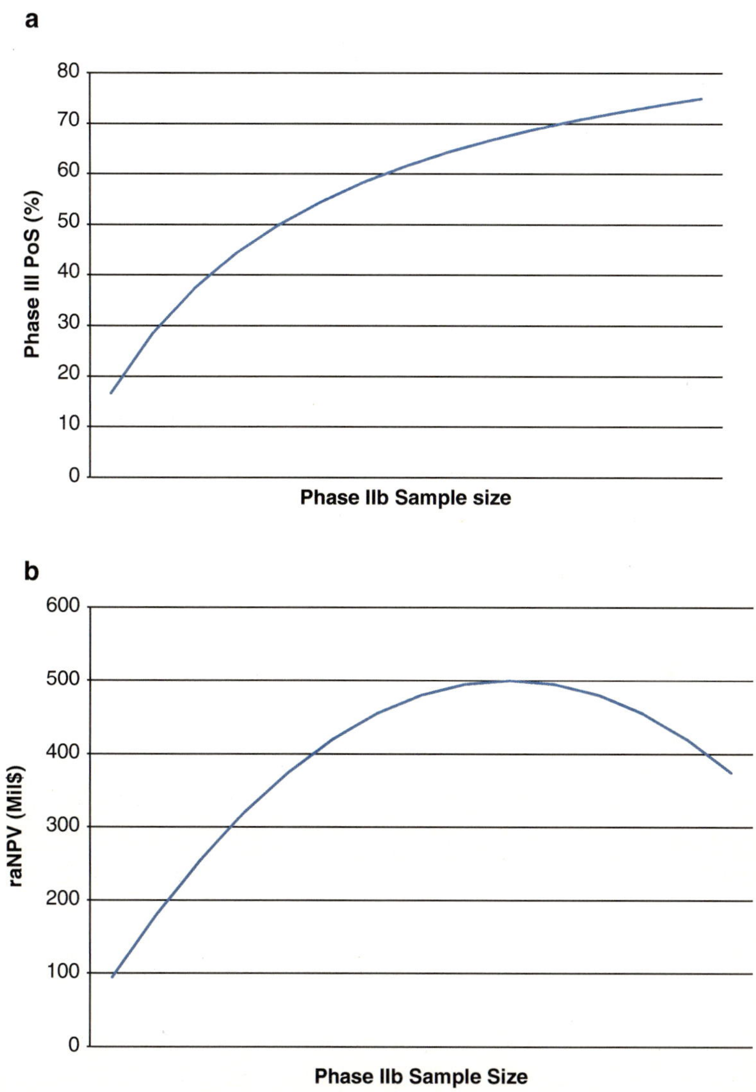

Fig. 1.1 PoS and expected NPV relationship to sample size. (**a**) Relationship between the size of phase IIb and the probability of success in phase III. (**b**) Relationship between the size of phase IIb and product's expected NPV. *Source*: Reproduced with permission [8]

pharmaceutical portfolios, but have explained in the introduction as a mean of avoiding repetition, since several chapters refer to ENPV. Other measures, such as the benefit–cost ratio (BCR) are more robust to these uncertainties, at the sacrifice of details [7]. Several chapters in this book are actually using the BCR as the measurement of interest.

Value of Design in Decision Making

The current assessment of PoS is often based on high-level understanding of drug development. Parameters that are considered are the indication, and the stage of development. Based on this limited information estimated PoS' are plotted against expected revenues, and this is information is then used to support decision making.

This approach completely overlooks the value of design. Multiple development options (e.g., doses, designs, endpoints, budget constraints) should be compared based on the expected utility, as the expected value of a product clearly depends not only on the quality of the product itself but also on the quality of the development program. Therefore, study design and development strategy should always be considered when making decisions.

Let us illustrate this with one example of impact of design on the value of the product, at the program level. Antonijevic et al. [4] studied the impact of Phase 2 design characteristics on the PoS in Phase 3 and on the ENPV of the product. The following Phase 2 characteristics were studied: (1) the statistical approach to dose selection, (2) the sample size used in Phase 2, (3) the number of doses studied in Phase 2, and (4) the number of doses selected to advance into Phase 3.

There were seven different statistical approaches considered, and they are presented on the *y*-axis in Fig. 1.2. Two different sample sizes were considered in the

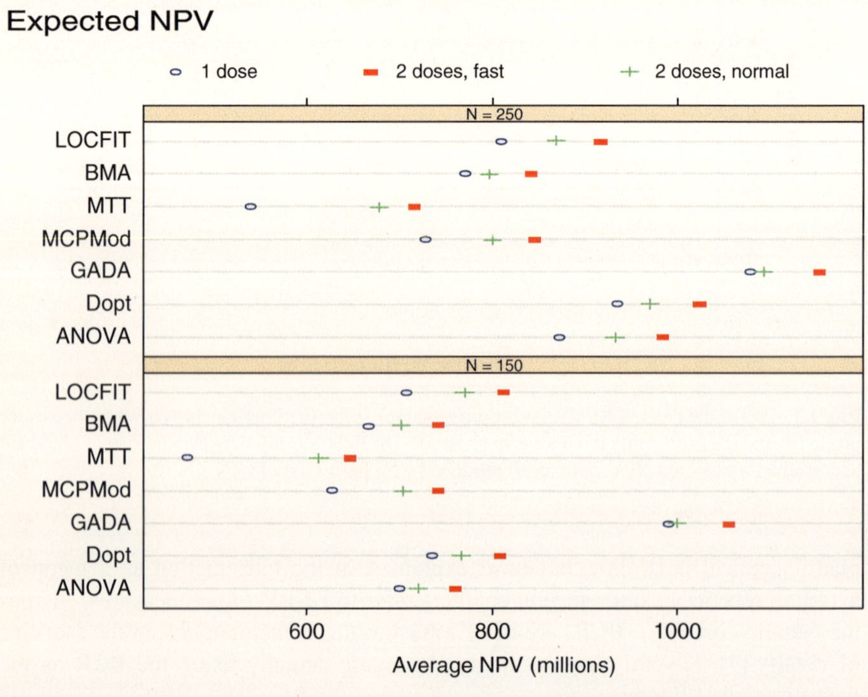

Fig. 1.2 Illustration of value of design. *Source*: Reproduced with permission [4]

Phase 2 (150 and 250), and three options for the Phase 3 (one dose, two doses, and two doses with accelerated enrollment). This results in the total of 42 Phase 2b/III strategy options to be compared. Results of this study are discussed in more detail in the chapter on Adaptive Programs, and only the point regarding the value of design is made here. The PoS for selected development options were estimated using simulations. Costs and development times were calculated using the information from real databases. The resulting ENPV for selected design options ranges from approximately \$0.5B to \$1.2B. This is the range of the value of design. Definitely something that should not be overlooked in planning and decision making.

Uncertainty

As previously mentioned, optimization of ENPV requires addressing uncertainties of the development process, while updating the incremental knowledge as one progresses down the drug development life cycles. Knowledge acquired in the process of developing a drug may also alter PoS for other follow on drugs that work in the same pathway [1]. In these settings, Bayesian statistical approaches are a natural tool for combining various sources of prior information. If very little prior information is available at a given stage of development, then one can consider a Bayesian approach where industry averages and/or clinical opinion are used as priors. The updates can then be made by using the observed data from ongoing trials. One approach to addressing uncertainty in development programs is to adopt a concept of assurance, or unconditional probability of a positive outcome [9]. Here "unconditional" dedifferentiates from the traditional approach of calculating power, which is "conditional" on prespecified assumptions. The idea of computing the expected (or "average") power with respect to the prior distribution of the parameter of interest is not a new concept; however, it has rarely been implemented when planning new studies. For detailed discussion of a combined Bayesian and frequentist approach to study design as a mean of incorporating unconditional PoS see Spiegelhalter et al. [10, 11].

Making Decisions in the Broader Context

Optimization at the Portfolio Level

So far we discussed the value of a product, and optimization at a development program level. It is now time that we address our main topic: Portfolio Optimization. A pharmaceutical portfolio will include multiple products, candidate predictive biomarkers, and clinical trials. Assessing the value of a portfolio therefore includes all the parameters described previously, but also requires additional considerations. First, budget limits are set not at the product level, but at the portfolio level. This makes any decisions interrelated and increases the complexity of decision-making.

For example, reduction of costs in one program does not only mean immediate savings but also enables increase in investment in other programs.

Further, given the budget constraints, not all planned programs and clinical trials can be executed. One needs to focus on the programs expected to bring the greatest returns, which raises the question of project selection. Burman et al. [12] proposed a solution to this problem by a decision-tree approach. The resolution to the budget constraints problem, however, should not only be in deciding which projects to select but also how large they should be. It is standard practice in the industry to fix the sample size of Phase 3 studies based on a predetermined level of power (usually 0.80 or 0.90). In this way, a study sample size is not explicitly linked to the commercial value. We argue, on the other hand, that the optimal portfolio development solution should be a combination of project and sample size selections, so that the value of a portfolio as a whole is maximized.

Finally, portfolio optimization cannot be static. Sponsors of Phase 3 trials are most often large or medium sized companies with pipelines of drugs at various stages of development. Phase 3 portfolio development strategy will thus require planning over a time horizon within which a number of viable candidates are expected to become available at various times. Furthermore, many of the potential Phase 3 drugs will have uncertainty surrounding their availability because they will be in earlier stages of development and may fail to progress to Phase 3. Portfolio planning is also affected by external dynamics. Events, such as approvals of new drugs, would have a major impact on the expected revenues and as such affect the ENPV. Dynamic optimization of the implementation schedules and reoptimization in the light of accumulating information is therefore a key requirement for the Phase 3 portfolio optimization. There is thus a challenge in keeping all the important variables up to date so that each decision can be informed by the context of available portfolio information.

Project Prioritization vs. Portfolio Optimization

Project Prioritization often gets mistaken for Portfolio Optimization. Let us illustrate the difference by looking at sample size in Phase 3 as the parameter of interest. Any portfolio would include a number of products or development programs. For each product the function of sample size vs. ENPV would be as illustrated in Fig. 1.1b; the only difference being the slope of the curve, and the location of the inflection point. These differences would be a result of different costs, expected revenues, and chances of success. Then let us assume that we have a certain budget limit set at the portfolio level.

1. Strategy 1, Project Prioritization: Determine sample size for each trial and calculate the ENPV. Then do a naive selection among trials with highest ENPV to fit within budget limits.
2. Strategy 2, Portfolio Optimization: Start all trials with sample size = 0. Compare trials for the benefit gained from an increase in sample size in fixed increments, say ten patients. In each iteration increase the sample size for trial with that has the steepest slope on the ENPV curve. Repeat this procedure until the budget limit is met.

Clearly, the second strategy will always perform better than the first strategy. The key difference between two approaches is that optimization considers study design as a variable that can be manipulated such that the value of a portfolio is maximized. Project prioritization assumes that study design is fixed, and as such results in suboptimal outputs.

Need for an Integrated Input by R & D and Commercial Groups

In this section we briefly describe why the integrated approach is necessary in order to maximize the value of Pharmaceutical Portfolios. We first address this from the commercial point of view, then from that of R & D.

Commercial View

There is an increasing demand for differentiated product value. There is also a growing recognition that the traditional, *siloed* organizational focus on product development separate from launch and beyond will no longer meet post-launch product needs. As a result, an integrated approach is needed that considers value proposition earlier in the clinical development process by incorporating marketing, commercial, and medical affairs perspectives. Further, the R & D teams cannot provide meaningful solutions without understanding the external environment, including status of any competing products. Finally, drug approval is of little meaning if reimbursement from payors is not forthcoming.

R & D View

Commercial decisions are based on the top level understanding of drug development. It is the R & D team that has deep understanding of the data collected, regulatory and clinical strategies, different development options and their impact on the approval process. Commercial teams have limited understanding of drug development risks and uncertainties.

Examples Where an Integrated Approach is Necessary

Dose Selection. A well selected dose with an optimal safety/efficacy/health outcomes profile would impact many parameters that we discussed before. It would improve chances of regulatory success and reimbursement, and would also potentially increase the market share. In order to improve the chance of selecting such a dose, however, one needs to invest into a more robust Phase 2 that would increase the cost, and lengthen the time of development. For antibody therapeutics, it may be desirable to use labeled biodistribution studies in conjunction with Phase 1 work for the purposes of dose determination [13]. Finding an optimal solution for this problem can be done only with joint input from R & D and commercial.

Program-level optimization. Another example, and a much broader one, is the optimization. Calculating sample sizes has been one of the main tasks for statisticians. Financial outcomes and optimization, however, have rarely been considered when sample size assumptions are selected. During Phase 3 planning, power is usually arbitrarily prespecified, without assessing the extent by which the ENPV would be affected by changes in sample sizes, or what would be the "optimal" power. Likewise, proof-of-concept (PoC) decisions and interim decision rules are rarely set such that the expected revenues can be optimized. It should be noted that optimal sample sizes would be classified by conventional statisticians as "underpowered" in many cases [1, 7].

Portfolio-level optimization. The optimization of various decision parameters can be done at the program level, but is better assessed in the context of a portfolio. Consider a portfolio with several clinical trials addressing different indications for which the expected costs and revenues largely differ. Should the power for Phase 3 trials be equal for these programs? Shouldn't the decision criteria following interim analyses be based on expected financial losses and gains, and ultimately be specified such that the expected value of the portfolio be maximized? For each drug program, there may be a portfolio of candidate predictive biomarkers as well, leading to a larger an even larger variety of portfolio options [1]. Please see Patel and Ankolekar [6] and Chen and Beckman [7] for an excellent discussion.

Summary

Numerous economic factors are important in the planning, design, and management of pharmaceutical programs and portfolios. In drug development decisions regarding various development options should be defined such that the expected value of a product or a portfolio is maximized. This should be done through the process of quantitative optimization of development decisions. While this optimization could focus on individual development programs, it is preferable that these decisions are made at the portfolio level, as individual programs are interdependent. Integration of various sources of information and input is essential in order to assure successful delivery.

Key concepts that are addressed in this book are as follows:

1. Drug development decisions should be made at the portfolio level. Budgets are limited and are determined at the portfolio level; therefore decisions within individual programs should be interrelated.

2. The value of a product depends on quality of all of the following: product itself, associated predictive biomarkers, and the development program. The value of a portfolio will then depend on the expected value of individual products as well as the strategy for portfolio optimization. Study design has a major impact on the success of drug development at:
   - *Trial level* by application of adaptive design. Examples are: early stopping for efficacy or futility, de-risking development by increase in power or interim analyses as needed, or adaptive population enrichment.
   - *Program level.* More effective dose-finding leads to higher success rates in Phase 3 and an improved efficacy/safety profile.
   - *Portfolio level.* Improved allocation of a fixed budget into individual trials leads to an improved value of the portfolio.
3. An integrated, R & D and commercial approach is necessary for optimizing pharmaceutical portfolios.

Contents and Organization of this Book

In this book we first present organizational and financial aspects of portfolio optimization. First chapter highlights clinical and other important aspects of risks associated with pharmaceutical product portfolio management, including competitive intelligence (CI), due diligence (to acquire assets and estimate the probability of technical success), and patent protection and regulatory exclusivity. The following two chapters discuss financial and business considerations when investments are made in a portfolio of pharmaceutical products, including how introducing flexibility in study design can de-risk investments and maximize the value of pharmaceutical products and portfolios where outside investments were made. The last chapter in this section describes challenges of portfolio management in "Big Pharma." It contrasts two commonly accepted approaches: a project-driven approach to a portfolio-driven approach, and then describes a third, compromising alternative.

Second part of the book describes new methodologies for optimizing pharmaceutical development programs and portfolios. This part of the book begins with a chapter that discusses optimization of drug development programs, which is a step towards optimizing portfolios. This is followed by two chapters with application of decision analysis: one to support the design of clinical trials at a program level, and the other that compares three different approaches to indication sequencing. Then follows is a work describing specific methods that maximize return on investment at the PoC stage, both at the program and at the portfolio level. The fifth methodological chapter is on optimization of portfolios that include development of predictive diagnostic biomarkers. Finally, a mathematical approach to dynamic portfolio optimization due to changes in the internal and external environment over time is presented.

References

1. Beckman RA, Clark J, Chen C (2011) Integrating predictive biomarkers and classifiers into oncology clinical development programmes. Nat Rev Drug Discov 10:735–749
2. Burman CF, Hamren B, Olsson P (2005) Modeling and simulation to improve decision-making in clinical development. Pharm Stat 4:47–58
3. Nixon RM et al (2009) The rheumatoid arthritis drug development model: a case study in Bayesian clinical trial simulation. Pharm Stat 8:371–389
4. Antonijevic Z, Pinheiro J, Fardipour P, Lewis RJ (2010) Impact of dose selection strategies on the probability of success in phase 3. Stat Biopharm Res 2(4):469–486
5. Brigham EF, Houston JF (2009) Fundamentals of financial management. South Western College Publications, Boston, MA
6. Patel NR, Ankolekar S (2007) A Bayesian approach for incorporating economic factors in sample size and design for clinical trials and portfolios of drugs. Stat Med 26:4976–4988
7. Chen C, Beckman RA (2009) Optimal cost-effective designs of phase 2 proof of concept trials and associated Go-No Go decisions. J Biopharm Stat 19:424–436
8. Antonijevic Z (2009) Impact of dose selection strategies on the success of drug development programs. TouchBriefings (4):104–106
9. O'Hagan A, Stevens JW, Campbell MJ (2004) Assurance in clinical trial design. Pharm Stat 4:187–201
10. Spiegelhalter DJ, Frieedman LS (1986) A predictive approach to selecting the size of a clinical trial based on subjective clinical opinion. Stat Med 5:1–13
11. Spiegelhalter DJ, Abrams KR, Myles JP (2004) Bayesian approaches to clinical trials and health care evaluation. Wiley, Chichester, UK
12. Burman CF, Grieve AP, Senn S (2007) Decision analysis in drug development. In: Dmitrienko A, Chuang-Stein C, D'Agostino R (eds) Pharmaceutical statistics using SAS®: a practical guide. SAS Institute Inc., Cary, NC, pp 385–428
13. Gatica G, Papageorgiou LG, Shah N (2003) Capacity planning under uncertainty for the pharmaceutical industry. Chem Eng Res Des 81(6):665–678

Part II
Background

Chapter 2
Clinical Aspects of Pharmaceutical Portfolio Management

Frederic (Rick) Sax, Raymond A. Huml, and Judith Ng-Cashin

Introduction

This chapter examines the clinical aspects of pharmaceutical portfolio management from the perspective of an Executive Committee reporting on portfolio strategy to a Board of Directors.

Portfolio strategy decision-making is the single most important role of pharmaceutical executives interacting with the Board of Directors to drive shareholder value. The broad categories driving shareholder value include a thorough evaluation of manufacturing and clinical drug development costs (e.g., cost of goods [CoG]), commercial (sales) expectations, and the clinical risk–benefit assessment for each product within the portfolio.

This chapter highlights the clinical aspects of pharmaceutical portfolio management. It also discusses other important aspects of the risks associated with pharmaceutical product portfolio management, including competitive intelligence (CI), due diligence (to acquire assets and estimate the probability of technical success), patent protection and regulatory exclusivity.

For the purpose of this discussion, an "Executive Committee" is a collection of internal experts covering the disciplines of medicine and drug development sciences, regulatory science, commercialization, and finance, as they relate to portfolio development and asset prioritization. A "Board of Directors" is a body of elected or appointed members who oversee the activities of a company or organization with accountability for corporate governance, financial resources, including acquisition and allocation, and stakeholder accountability. In addition, for purposes of this

F. Sax, M.D. • R.A. Huml, M.S., D.V.M., R.A.C. (✉) • J. Ng-Cashin, M.D.
Quintiles Transnational Corp., 4820 Emperor Blvd, Durham, NC 27703, USA
e-mail: Rick.sax@quintiles.com; Raymond.huml@quintiles.com;
Judith.ng-cashin@quintiles.com

Z. Antonijevic (ed.), *Optimization of Pharmaceutical R&D Programs and Portfolios: Design and Investment Strategy*, DOI 10.1007/978-3-319-09075-7_2

chapter, a typical pharmaceutical portfolio contains 10–15 assets, representing some mix of small molecules and biologics, as well as pre-approval assets and marketed products.

The portfolio strategy and management is examined in the context of clinical discovery and development, acquisition of external assets, life cycle management of marketed products, risk–reward tolerance, and clinical risk mitigation strategies.

While this chapter will be most applicable to portfolios of small molecules, biologics, or medical devices, it could also apply to other diversification strategies for biopharmaceuticals portfolios, including moving into other sectors such as wound healing, consumer products, generics, and agricultural or veterinary products. Smaller companies may benefit from the strategies elucidated in this chapter, but may not have the infrastructure to diversify and balance risk in a similar manner due to the more limited nature of their portfolios and available resources.

Portfolio Strategy Decision-Making

Portfolio strategy decision-making focuses on five key elements of the clinical assessment which is divided into both strategic and tactical components.

The strategic clinical components include an understanding of the overall strategy (e.g., perhaps a focus on one therapeutic area or multiple areas or types of products, the ability to assess the probability of technical success and the value of the clinical and technical assessment of the portfolio which is a combination of clinical risk and NPV risk) to obtain alignment with the overall corporate strategy ("strategic fit") and priorities, and the value that is created from resources being appropriately deployed to maintain revenue from marketed and future products ("value creation"). Contained within this paradigm is the tension of maintaining a balance between current and future revenues through investments in the development pipeline.

The tactical components include the methods in the pharmaceutical executive's armamentarium to assess risk. Clinical risk is assessed by known safety signals as well as unanticipated safety signals which could be seen after a product is marketed. Risk can be managed by due diligence, assessing the risk associated with in-licensed compounds as well as ongoing due diligence on the compounds in the portfolio. Other tactical components include employing methodologies to avoid bias with a careful consideration of the trade-offs for each decision in order to optimize the ever-changing portfolio.

The clinical contribution involves product development risk (but not patent risk) and estimating the probability of technical success (PoTS), which includes an evaluation of safety risk, probability of regulatory success, and market access. To avoid bias, due diligence must be conducted both for in-licensed assets and, on an ongoing basis, for the existing portfolio.

Depending on the clinical mix of the portfolio, as determined by the therapeutic areas, the value of the entire portfolio then becomes a combination of clinical risk and net present value (NPV) risk.

The strategic and tactical risks must be managed in a sustainable manner, meaning that the goals must be achievable with the available resources, and the returns appropriate to the level of investment, all in alignment with the growth objective.

Portfolio management must strike an appropriate balance between all of these factors.

Strategic Fit

The portfolio must be aligned with the overall corporate strategy and priorities. This includes both current and future strategies. For instance, if the early phase pipeline is weak and there is an aspiration to strengthen it in the near term, which strategy is more consistent with the future strategy: hiring in more expertise within a key discovery discipline or in-licensing early phase assets with the appropriate profile from external organizations? In other words, build it or buy it? The tactic taken here will certainly be determined by the infrastructure strategy, the importance of having internal expertise/capability for the future portfolio, and the cost structure associated with either course of action.

Value Creation

It is the primary goal of portfolio management to maximize value and return. Several considerations must come into play when addressing the composition of the portfolio toward this goal. The short term revenue goals of the portfolio can take priority over investment in the development pipeline. However, inattention to the future marketed product strategy enabled by assets in discovery and development can result in devaluation of long-term future revenue potential. Determining the optimal resource allocation toward sales and marketing for current revenue maximization versus investment in research and development of the less mature asset pipeline can be difficult and must be aligned with the strategic goals of the company.

Balance

The composition of the portfolio must be balanced and commensurate with the financial goals, corporate strategy, and risk tolerance of the company. The Executive Committee must carefully consider many parameters, and then make thoughtful and often difficult decisions about which product(s) to include in or exclude from the portfolio.

For pipeline assets, often the risk of a project is proportional to the projected return on investment. When thinking about pipeline opportunities, it is often the high risk-high return projects that also require the most internal resource (see Fig. 2.1).

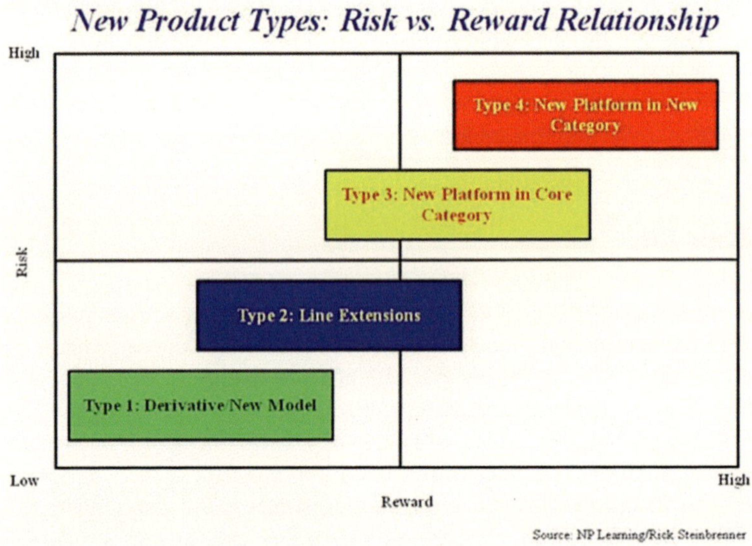

Fig. 2.1 Risk vs reward relationship for four different types of new products

Both risk tolerance and consideration of consequences of resource demand and reallocation should be considered. For example, if a high risk-high return project consumes the high performing talent, what low-medium risk but more certain revenue return projects will fall behind? A portfolio dominated by early phase assets might have a high potential value, but will also carry a high attrition rate, while a portfolio containing mostly marketed products may encounter a revenue "cliff" without revitalization from the pipeline.

Tactics for Clinical Portfolio Management

Complete analyses of tactics and strategies for continued development or termination of an asset, or inclusion, exclusion or abandonment of a product—whether in the preregistration or post registration phase of the drug cycle—is beyond the scope of this chapter. However, in terms of clinical decision-making, the product or potential product must have an acceptable risk–benefit profile for the specific indication in the appropriate patient population and must be adequately differentiated from the existing available therapies. The risk–benefit profile will be informed by severity of indication, magnitude of medical need, and local standard of care. A detailed clinical evaluation must be performed and continually reassessed at key investment milestones in order to ensure informed decision-making for portfolio inclusion.

Clinical Considerations

The best person to evaluate the practicality and clinical use of a medicinal product is a physician with clinical experience in the therapeutic area or indication for the product. The physician can provide unique insights such as:

- Ease of use.
- Potential for common drug–drug or drug–disease interactions.
- Patient population.
- Metabolic effects.
- Adverse effects common to the class of drug.
- Competitive environment.
- Clinical need for a new product.

This clinical review may also have implications from a commercialization standpoint.

The key clinical consideration is how the product addresses the unmet medical need. Another major purpose of the clinical review is to determine the safety and efficacy profile of the product—in terms of the product itself and the portfolio—as defined within the confines of a risk–benefit assessment. This assessment is easier for approved products or for a known class of products, and more difficult for new chemical entities or those entities that lack clinical data (e.g., preclinical candidates). Efficacy in animal models does not always translate to humans and thus care must be exercised—and higher risk ascribed—in the evaluation of earlier stage products. Other clinical considerations for asset/product assessment include, but are not limited to:

- Is the product unique? If not, does it complement or compete with the existing portfolio?
- How easy is the product to use?
- Can the product be sold using the existing sales force or will it require additional resources such as a specialty sales force?
- Doe the product require extensive training of physicians (e.g., certain invasive cardiovascular medical devices) or patients (e.g., self-administered hemophilia product) or no training at all (e.g., OTC)?

Risk Management

Overall

Risk must be managed to protect a 10-year time horizon, which involves considerable uncertainty, in the context of current and anticipated investments within an ever-changing competitive environment.

Two key determinants for product risk management are patient risk, defined here as an evaluation of safety, and product development risk, which includes a risk assessment of clinical, regulatory and unknown product safety after approval but prior to patent expiration. Other determinants that will be discussed further include protecting existing compounds, in-licensing, mergers and acquisitions, due diligence, competitive intelligence, and patent protection and regulatory exclusivity.

Protecting Existing Compounds

There are a variety of R & D strategies to maximize or protect the value and return of a marketed product portfolio. For individual assets:

- Improving dosing schedule and patient convenience through formulation development.
  - Development of an extended release formulation from an existing immediate release formulation.
  - Improving patient palatability through improved dose presentation.
- Improvement of safety or tolerability through formulation/excipient modification.
  - Liposomal formulation to mitigate renal toxicity.
- Supplemental application for new indications for a marketed product.
  - Related indication within therapeutic area (such as a renal carcinoma application following melanoma approval).
  - New application in a dissimilar therapeutic area (e.g., an autoimmune disease application following cancer approval).
  - New application for an Orphan drug indication (this strategy also garners certain extra regulatory exclusivity that can act as patent protection).
- Approval in a new geography, region, or country.
  - Including emerging markets.
- New patent, extension of existing patent, or additional protection in a certain geography.
  - Pediatric clinical studies in response to an FDA written request (pediatric exclusivity).

In-Licensing, Mergers, and Acquisitions

If the R & D pipeline is limited, external sources of complementary (and also non-competing) products—which must meet an unmet medical need or be differentiated from the existing marketed products—should be examined to create value. Sometimes, for specific needs, it may be necessary to in-license a late-phase opportunity in a nonstrategic area, simply to maintain cash flow. If the need is considered great enough, and capital is available, an entire company can be purchased which might include a specific compound (or compounds) or perhaps a platform technology that may either produce additional candidates or modify existing compounds in the buyer's pipeline.

One possible response may be to in-license a product similar to an existing, successful marketed product with the goal of adding (or replacing) market share from the existing product. However, the new product must be sufficiently differentiated in order to achieve regulatory approval and payer reimbursement.

Below are two case studies which highlight the success of second generation, follow-on prescription drug (Nexium®) and a patent dispute designed to ward off generic competition (Alimta®).

Case Study: AstraZeneca's Prilosec and Nexium

AstraZeneca had a dominant gastrointestinal franchise with Prilosec® (omeprazole), which was the world's best-selling drug, with global sales of $6.2 billion. To counter the potential downside of Prilosec's patent expiry in October 2001, AZ CEO Tom McKillop considered a range of options, according to a *Harvard Business Review* Case Study titled "AstraZeneca, Prilosec®, and Nexium®: Marketing Challenges in the Launch of a Second-Generation Drug," by James G. Conley, Robert C. Wolcott, and Eric Wong (January 1, 2006). These included several "franchise-extending" strategies, such as the launch of a second generation, follow-on prescription drug (with a new patent) branded as Nexium®, and the introduction by AZ of generic omeprazole and/or an over-the-counter version of omeprazole. The *HBR* case study notes that both the generic and OTC markets were uncharted territory for AZ. "The path forward to sustain market dominance in its category, especially with respect to the OTC opportunity, would require significant channel know-how." Clearly, AZ successfully tackled this challenge, successfully launching Nexium®—with a differentiated product profile—to protect its income stream after Prilosec's patent expiry. Nexium® ranked second in sales in the US market in 2011, with sales of $6.3 billion according to IMS Health [1]. In the second quarter of 2012, Nexium® reportedly led the US market with sales of $1.38 billion [2].

Case Study: Lilly's Unusual Patent Dispute

In August 2013, Eli Lilly & Co. began defending a patent for its lung-cancer treatment Alimta®, which recorded global 2012 sales of $2.6 billion, according to a *Wall Street Journal* report [3]. The patent covers the method of administering Alimta® to patients with certain vitamins designed to mitigate side effects. A different patent covers the basic chemical composition of Alimta®. The newspaper writes that the case highlights the pressure on drug makers to preserve market exclusivity for top-selling products for as long as possible in the face of generic competition, pricing pressure and underproductive research labs. A victory for Lilly would block several generic companies from selling low-cost copies of Alimta® in the US market until at least 2,022, notes the *WSJ*. A loss could allow generics to be launched in 2017—when the patent expires for the basic compound—a development that would rapidly erode sales of the original brand.

Due Diligence [4]

For purposes of this chapter, due diligence (DD) is used to assess the probability of technical success (PoTS). Technical risk is defined as those factors that are inherent in the product and will contribute to its full sales potential, given the right sponsor.

Due diligence is simply a process for managing risk. All companies perform DD prior to making an investment. Proprietary information is first exchanged between companies after a Confidential Disclosure Agreement (CDA) has been executed. This is of critical importance to protect both parties and should be executed promptly, usually under the direction of a company's legal department/counsel.

Environments in which DD can be utilized range from simple, single-product transactions between a buyer and a seller to more sophisticated global acquisitions of multiple products. The simplest DD exercise may require only one person; more sophisticated partnering opportunities or acquisitions may require a team of experts with a range of disciplines. Due diligence proceeds with this team of experts to assess corporate strategy, research and development, intellectual property, human resources, and financial dealings, identifying the strong points and weak points of a company, a product (or products), or even a potential deal in order to better manage risk.

Competitive Intelligence [5]

Competitive Intelligence (CI), for purposes of this chapter, is used to determine market risk while identifying competitive threats so that they can be addressed as early as possible.

Pharmaceutical CI entails defining, gathering, analyzing, and distributing intelligence—both nonproprietary and proprietary—on pharmaceutical products, customers, competitors and any aspect of a particular functional area needed to support executives and managers in making strategic decisions for an organization (e.g., an expected return on investment or strategies based on the loss of patent protection).

Stakeholders are varied and include pharmaceutical companies, contract research organizations (CROs), pharmaceutical manufacturers and those associated with the supply chain, investors, patients, health payers, and government organizations. Although typically thought of as being driven by other companies, competition may also be affected by regulations (including product-based labeling), lack of regulations (e.g., lack of Guidance from FDA's Office of Prescription Drug Promotion

(OPDP) regarding social media) or long-awaited draft FDA guidance for biosimilars, finally issued in February 2012, politics (e.g., controversy around medical cannabis-derived products), accounting principles (e.g., general accepted accounting practice (GAAP), geographies (International Conference on Harmonization (ICH) vs. non-ICH), patent protection, and regulatory exclusivity). Patent protection and regulatory exclusivity is discussed separately.

Publically available information is obtained via the World Wide Web and may be accessed for free, such as information contained on a competitor's Web site, or available for a cost (e.g., fee to print a full-text article or IMS Health data to track pharmaceutical sales). It is typically limited by the savvy of the researcher, the amount of time that the investigator has to compile the information, and by the investigator's access to company-wide databases. Large companies typically have an advantage over smaller companies in gathering CI due to their scale and resource availability.

A key caveat to this entire process is that the gathered information must be converted into intelligence and then utilized for business decision-making when assessing the market for a particular product or group of products. In essence, if the CI gathered is not usable (or actionable) then it is not intelligence. Increasingly important is the understanding of the landscape for the payer environment when assessing the overall risk of any particular product.

Patent Protection and Regulatory Exclusivity

The available patent life or regulatory exclusivity of any product must be taken into account when determining the return on investment. This is even more important for portfolios of products, such as biosimilars, where the originator company is standing ready to battle potential future competitors. Therefore, it is imperative that the pharmaceutical executive have a general understanding of the overall process for an assessment of patent protection and regulatory exclusivity.

The strength of a patent, its remaining patent life, and the potential to obtain regulatory exclusivity all form the basis of protection for a branded product from competition, including generics. Another key issue is freedom to sell a product without interference from third parties that may own relevant patents. Because pharmaceutical executives are increasingly being asked to participate on due diligence teams, they need to be familiar with the IP investigational process and the key outputs of the IP assessment. This enables their employers to better understand the risks associated with the inevitable patent challenges that arise with financially successful branded products and potential threats from third-party patent owners.

Although respected in major ICH countries (e.g., the USA, EU, and Japan), not all countries honor patent protection equally and this reality must be factored into a global marketing strategy. Moreover, the patent and regulatory exclusivity situation for a given product often varies substantially in the various countries. For sponsors of branded products, this may preclude marketing a product in a particular country.

Key considerations for understanding the value of a product relative to patent protection and regulatory exclusivity include:

- IP assessment is critical to the entire due diligence and product portfolio optimization process.
- IP protection includes both the classical IP assessment as well as the regulatory exclusivity assessment. Together they form the basis of protection for a product from competition.
- When IP vulnerability is discovered, it is often possible to diminish or even eliminate a risk through contractual language.

Avoidance of Bias

Overall

There are multiple ways to reduce or avoid the biases inherent within product portfolio management. This section provides an overview of the pitfalls associated with potential biases in decision-making and discusses several successful approaches for reducing risk.

Background

When new products come out of the discovery phase of drug development, they typically look very promising and commercial hopes are high. Most candidates, nonetheless, fail during Phase I or Phase II.

Development teams often work in "plan-to-succeed" mode, without necessarily performing war-gaming exercises to predict what might go wrong, and without adequately challenging the probability of scientific, regulatory, or commercial success as anticipated by the due diligence or product marketing teams.

War-gaming can help determine whether a portfolio plan is truly robust, using measures of the actual risk within the portfolio—clinical, regulatory and commercial—to calculate an Expected Net Present Value (ENPV, which takes into account estimated probability of success for each product under development). This metric is more useful than an NPV (which assumes 100 % probability of success) and can be risk adjusted to take into account the probability of success for compounds in the pre-registration phase of drug development. While some risks are known—for example, the probability of patent expiry is 100 %—others are less certain, such as the:

- Probability that the animal data predicts clinical outcomes.
- Probability of the occurrence of unexpected safety findings.
- Probability that a product is adequately differentiated.

- Probability and timing of competitive entries along with their level of differentiation.
- Clarity of the path to regulatory approval and market access.
- Likelihood of success for the proposed marketing and sales approach.
- Political landscape (which may favor bringing cheaper copies of products to market, such as generics or biosimilars).

These factors and the assumptions around them can be used to collectively summarize the positive and negative attributes of each product in the portfolio. This, in turn, can be used to risk-adjust the overall probabilities of success for these products, thereby giving a more meaningful assessment of individual estimated NPVs and the net value inherent in the overall portfolio.

Predictive Tools

Once the known data about the product (or the scientific class of the product) are collected—and the scientific, regulatory and commercial assumptions about the product are clearly defined and delineated—it becomes possible to model the impact of this information on the probability of successful development and commercialization of the product.

While commercial modeling is a standard process for most pharmaceutical companies, modeling of potential scientific and operational development outcomes is rarely performed. More recently, a number of companies have developed software tools to better analyze and predict these development outcomes. Estimation of enrollment, overall development time (based on critical path activities) and cost of a program are examples of how tools can be used to estimate operational outcomes. Other technology, such as Quintiles Infosario Design®, use large, patient de-identified datasets (including electronic health records), to better understand population dynamics (impact on inclusion/exclusion criteria, accessibility of patients, standard of care, clinical event rates, etc.) and their impact on the design of clinical trials and programs. Use of such data, combined with existing clinical trial data, allows for the development of combined scientific and operational scenarios—which can then inform programmatic decision-making.

Access to such data and the tools to "prototype" outcomes by developing design scenarios, empower drug developers with the knowledge locked in such data and improve decision-making. Ultimately, better design decisions by minimizing bias and unchallenged assumptions, should help to improve the probability of success at each step in a product's life cycle.

Once drug development reaches trials in patients, these systems can be deployed as technology-assisted consulting services designed to help pharmaceutical companies optimize their clinical planning and design process. This optimization is achieved through two major steps: first, expanding the design space by generating new options and new combinations of approaches to the problem; and then optimizing time, cost, and risk parameters over the expanded space of design possibilities [6].

Adaptive Design Principles

Adaptive design principles—which address the risk embraced by both R & D and commercial factors—can be used to mitigate portfolio risk. Use of adaptive principles is an extension of the "cone of uncertainty" concept: As uncertainty is reduced when R & D milestones are achieved, or increased (e.g., due to lack of differentiation or unexplained safety events), the overall portfolio probability of success is modified as it "adapts" to the inclusion of this new information. Rather than iterative, isolated decisions are being made at a single individual product event, the overall portfolio balance is maintained, which minimizes bias, and accounts for the ever-evolving ENPV. This approach contrasts with the timing of most overall portfolio reviews performed by large biopharmaceutical companies that only may occur two or three times a year.

For a full portfolio analysis, the "cone of uncertainty" can be applied across all 10–15 molecules. Although potentially time consuming, each molecule has a risk adjusted ENPV, which changes at each R & D milestone. Adaptive statistics are applied across all "cones of uncertainty," taking into account variability in the confidence that can be attributed to the ENPV. This process has the important benefits of being more methodical and less subject to bias than traditional, infrequent portfolio reviews.

Aspirational Drug Development

Other biases that can corrupt the portfolio include teams engaging in "aspirational" drug development. In this case, the drug development team aspires to an outcome—that may or may not be grounded in the scientific realities of the drug's safety and efficacy performance. For example, a marketing team may have the desire to have a safer drug, but the clinical data may not demonstrate a difference in adverse events (AEs) using descriptive statistics, or the drug may not have even been studied in such a way that a superiority safety claim can be justified. Since the sales team is limited, from legal and ethical standpoints, to discussing only those claims that are justified in the approved label, this leads to a mismatch in the desired "target product profile" and the "actual product profile" (as manifested in the approved label) and with corresponding impact on the commercial forecast. On the flip side, if designs of clinical trials and programs are based solely on the desired attributes of the drug rather than the manifest scientific "actual" attributes of the drugs, there is a high probability that the trials will fail to deliver against expectations, and thus generate little to no value for the product.

Increased or Decreased Chances of Success

Some situations may warrant optimism when determining the chances of product's success either in the clinic or the market. Well understood drugs, with a strong track record of historic sales, for example, should be expected to sell well when positive data is generated in the pediatric setting. An increase in the PoS for registration should also be expected for products with extended release formulations or 505(b)2 applications. The converse is also true. For example, some drugs that work in some settings may not work well in adjunct but apparently similar indications (at least from a therapeutic class standpoint), for example a schizophrenia drug moving into manic depressive disease.

Sustainability

The portfolio must be able to deliver on its short term goals with appropriately allocated resources and ensure a sustainable stream of revenue to continue to fuel product development and sales and marketing support. For any asset, competitive strength versus market attractiveness should be considered. Competitive strength is reflected based on a composite of market share, size/scale, quality, technology, cost of goods, brand strength, and customer loyalty to the class. Market attractiveness is a composite of size, growth, competitive rivalry, profit levels, and ability to differentiate.

While these parameters will be data-based for marketed products, modeling with careful assessment of the underlying market assumptions will need to be employed to assess this rubric for assets in earlier phases of development. Keen focus on the optimal target product profile is critical to ensure differentiation and reimbursement. To this end, the development plan must be designed such that robust data informing clear "go/no go" decisions is available at appropriate stage gates. These decisions, especially when the data support termination of development, can be difficult for organizations. However, rigor around this process is critical to make the best use of limited resources.

Portfolio Growth

Here, the goal is not only to protect but to increase revenues. Revenues can be obtained through strategic in-licensing (utilizing due diligence teams) or through desperation in-licensing. The latter may be due in part to known confounders, such as patent expiration. It may also be a response to unknown events such as an untoward safety signal (e.g., immunogenicity) that arises in clinical development of a company's own product or a competitor's, or a change in the regulatory climate (lack of FDA guidance or additional restrictions in a therapeutic area [e.g., cardiac outcome investigations with new compounds used to treat diabetes mellitus]). Some franchises may prove to be a scientific dead-end, with no new mechanisms-of-action coming out of R & D to create value. In that case, a company would need to create a new franchise, most likely adjacent to an existing, successful franchise, where the target market would be similar—for example, it might make sense for a company that is strong in diabetes to move into antiobesity products. Serendipity can also play a role in portfolio growth—for example, when scientists who are looking for activity in one therapeutic area find it in another (see Viagra case study).

Case Study: Viagra Discovery Based on Serendipity

"The surprising truth is drug design owes more to serendipity than careful design, and their potential may only be discovered when we take them," wrote *BBC.co.uk* on January 20, 2010 [7]. Quoting Viagra as an example, *BBC.co.uk* reported that this product, code named UK92480, stared life as a new treatment for angina. Trials in people were disappointing, and Pfizer was "about to abandon further trials when the trial volunteers started coming back and reporting an unusual side effect—lots of erections." Until Viagra's 1998 launch "there was no oral treatment for erectile dysfunction… Now, thanks to a failed angina treatment, men had another option. Viagra is now one of the most prescribed drugs in the world."

Summary

Portfolio strategy decision-making—taking into account development and manufacturing costs, commercial expectations, clinical risk–benefit assessments, competitive intelligence, due diligence, and intellectual property protection—is a key element in driving shareholder value.

Sustainable pharmaceutical portfolio management is guided by principles that maximize the value of the portfolio, balance its components, and make sure that additions and subtractions represent a strategic fit. Choices should be guided by an acceptable level of risk, as determined by the risk–benefit assessment for each product in the portfolio. The pharmaceutical physician is usually the best person to perform this review and make this assessment, although other persons with appropriate clinical and medical training can also conduct these assessments.

The key clinical decisions that must be made are based on an understanding of the pathophysiology of the compound, the safety and subject risk interpretation and the use of the product in the appropriate clinical setting (surgery suite, private practice, hospital, outpatient, etc.). Once the safety risks have been identified, they need to be managed within the bounds of regulatory affairs mandates (annual safety updates, annual reports, pharmacovigilance, etc.), ethical mandates, and best clinical judgment.

The portfolio must be balanced and driven by corporate financial goals, strategy, and risk tolerance. The portfolio must be able to deliver on its short term goals with the appropriately allocated resources to ensure a sustainable stream of revenue to continue to fuel product development and sales and marketing support. In addition to product development, a company can leap-frog this process and in-license a product or acquire a product or products via merger and acquisition.

When looking for products to acquire, it behooves the company to conduct due diligence and determine the positive, negative and unknown attributes of each product. The lynchpin of the due diligence process is the IP assessment which includes a thorough investigation of both patent protection and potential for regulatory exclusivity.

War-gaming is a useful approach in helping marketing teams determine whether the portfolio plan is truly robust, using measures of the actual risk within the portfolio—clinical, regulatory, and commercial—to calculate an Expected Net Present Value (ENPV).

Portfolio managers need to be aware of biases that can be inherent in portfolio decision-making, balancing those factors that may artificially increase or decrease the estimated probability of registration or commercial success. Complex tools, such as including adaptive design principles, and large data software tools, such as Infosario, can also be utilized to minimize and mitigate the risk and bias embraced by both the R & D and commercial factors.

References

1. Herper M (2011) The Best-Selling Drugs in America. Forbes, 19 Apr 2011. http://www.forbes.com/sites/matthewherper/2011/04/19/the-best-selling-drugs-in-america/. Accessed 24 Jul 2014
2. Staton T (2012) AZ's Nexium tops Q2 sales ranks, with BMS' Abilify on its heels. Forbes, 23 Aug 2012. http://www.fiercepharma.com/story/azs-nexium-tops-q2-sales-ranks-bms-abilify-its-heels/2012-08-23. Accessed 24 Jul 2014
3. Loftus P (2013) Eli Lilly enters unusual patent dispute. Wall Street Journal, 18 Aug 2013. http://online.wsj.com/article/SB10001424127887323455104579015291810290818.html
4. Huml RA (2010) Introduction to the due diligence process. RAPS, p 126, copyright 2010, ISBN: 0-9787006-4-3
5. Huml RA (2012) Pharmaceutical competitive intelligence for the regulatory affairs professional. Springer, p 82, Copyright 2012; ISBN: 978-1-4614-3681-2 (Paperback); ISBN 978-1-4614-3682-9 (eBook)
6. Quintiles press release (2013) Quintiles Infosario receives Computerworld Data+ Editor's Choice Award. 26 Aug 2013. http://www.quintiles.com/library/press-releases/quintiles-infosario-receives-computerworld-data-editors-choiceaward/. Accessed 24 Jul 2014
7. Jay E (2010) Viagra and other drugs discovered by accident. BBC News, 20 Jan 2010. http://news.bbc.co.uk/2/hi/health/8466118.stm. Accessed 24 Jul 2014

Chapter 3
Drug Development and the Cost of Capital

Kraig F. Schulz, Sarah T. Bobulsky, Frank S. David, Nitin R. Patel, and Zoran Antonijevic

The Situation: Pharma Industry is Struggling to Manage the Value of Equation

Regardless of industry or geography, the ultimate goal of almost any company is to create value. The definition and measurement of "value" may vary, but from a purely financial perspective, we can calculate the value of a company, product, or asset by present valuing the expected free cash flows using the relevant cost of capital. Failure to earn a return on invested capital in excess of the required cost of capital (and thus failure to create value) is, in fact, a failure of management, and often directly reflected in the market value of the company over time.

In the pharmaceutical industry today, overall value creation by R & D is, by many measures, essentially zero. The return on capital has been equivalent to (or sometimes less than) the cost of capital for most pharma companies. In fact, whatever "value" exists in the pharmaceutical industry is generated by on-market products,

---

K.F. Schulz, Ph.D. (✉)
Ernst and Young, LLC, 333 State St. #709, Bridgeport, CT 06604, USA
e-mail: kraigschulz69@gmail.com

S.T. Bobulsky
Ernst and Young, LLC, 5 Times Square, New York, NY 10036, USA

F.S. David, M.D., Ph.D.
Pharmagellan, LLC, 499 Adams Street, Box 94, Milton, MA 02186, USA

N.R. Patel, Ph.D.
Cytel Inc., 675 Massachusetts Ave, Cambridge, MA 02139, USA

Center for Biomedical Innovation, Massachusetts Institute of Technology, Cambridge, MA 02139, USA

Z. Antonijevic
Cytel Inc., 675 Massachusetts Ave, Cambridge, MA 02139, USA

Z. Antonijevic (ed.), *Optimization of Pharmaceutical R&D Programs and Portfolios: Design and Investment Strategy*, DOI 10.1007/978-3-319-09075-7_3

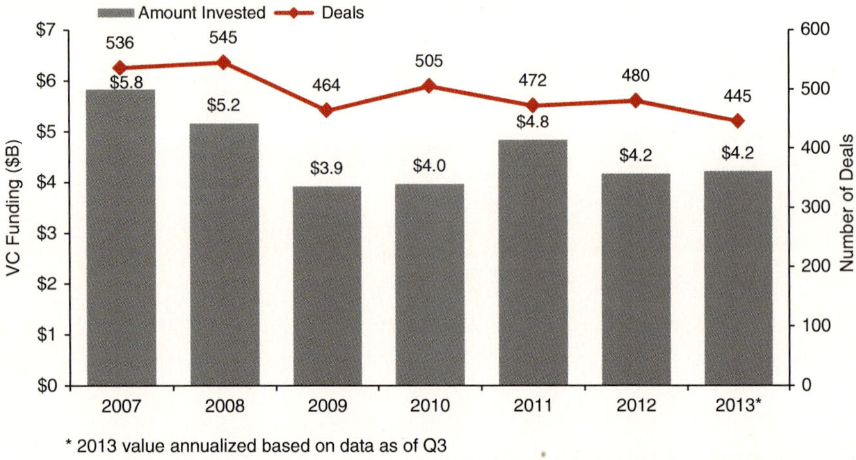

Fig. 3.1 VC funding of biotechnology (2007–2013). PricewaterhouseCoopers and the National Venture Capital Association's MoneyTree™ Report

with many companies' pipelines contributing no value, or worse yet, negative value to the valuation of the company. The cause of this is well-documented: products are too expensive to discover and develop, take too long to get to market, have dismal success rates in getting through development and face end-use markets that are under tremendous pricing pressure and significant competitive pressures.

One manifestation of this value challenge for R & D programs is the significant "bid/ask spread" between parties along the pharma R & D value chain. The difference between what an investor is willing to pay for the drug (i.e., the "bid") and the value of the drug as assessed by the developer (i.e., the "ask") results from differing perspectives on development costs, probabilities of success, time to market, and peak revenue potential differ—sometimes by orders of magnitude—between buyers (or financial sponsors like VC firms) and sellers. This discrepancy is reflected in a significant decline in early-stage pharma deals in the past few years. Despite the need for larger companies to bolster their pipelines via acquisitions of smaller biotechs or individual programs, the number of development-stage deals has declined. This is happening, in part, because venture capital firms, which typically fund early-stage R & D programs, are slowing their funding of biotech (Fig. 3.1). The same phenomenon is observed in big pharma R & D, with funding for early-stage investments being reduced in favor of later-stage assets. For both strategic and financial sponsors, the R & D process has failed to generate sufficient value for their investments. Put simply, the industry has not figured out a consistent way to manage the value equation of its R & D programs and, therefore, is seeing significant downward pressure on development-stage funding.

In light of this, there is a growing need for the pharma industry to more actively and effectively manage the R & D value equation by increasing the return on their capital, reducing their cost of capital, and/or generating revenues more rapidly as

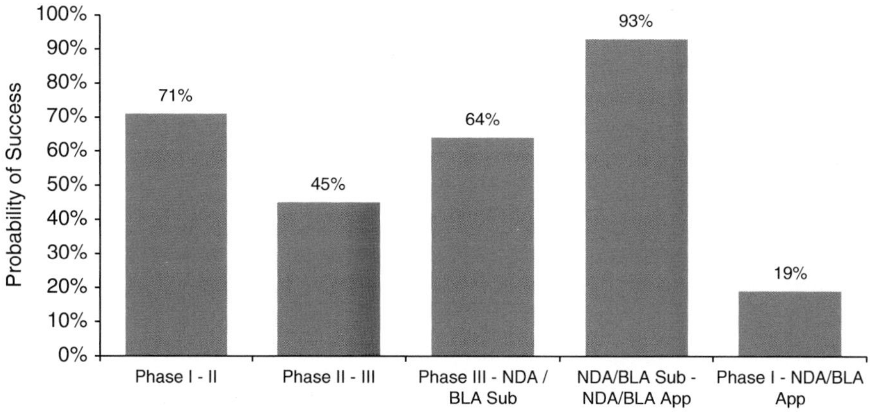

Fig. 3.2 Probability of success by phase. "Trends in Risks Associated With New Drug Development: Success Rates for Investigational Drugs," Nature Publishing Group, Volume 87, Number 3; DiMasi, Feldman, Seckler and Wilson; February 3, 2010

these are the key drivers of value. Given the historical risk–return profile of R & D in the industry vs. other investment alternatives, the cost of capital flowing into the sector is unlikely to get much cheaper. So how can pharma R & D improve returns on its invested capital? In theory, there are four "levers" that companies could use to improve R & D value: they could reduce risk by improving the probabilities of success (i.e., fewer trial failures), decrease trial costs (i.e., the amount of committed capital), shorten the trial and "decision-making" times (to reduce the duration of capital commitment and accelerate the realization of cash flows by getting the drug to market earlier), and/or seek to increase projected peak sales. Any of these adjustments could increase the value of a given program or the entire R & D portfolio.

In interviews with R & D executives, we find that pharma companies often focus on only one or two of these factors—decreasing risk or costs—in an attempt to deliver R & D value. These are often seen as the most "manageable" levers, though as we will see, pharma R & D executives' attention to these at the expense of others may be misplaced.

We begin by looking at risk. The inherent probability of success for any given molecule is unknown (Fig. 3.2). Furthermore, even if the molecule is *actually* effective, clinical trials may not demonstrate it as such. The probability of success of a molecule is therefore determined by its efficacy relative to the standard of care or placebo, and the power of the trial relative to the efficacy observed in the trial's sample patient population. The product's inherent efficacy is the ultimate unknown in the R & D process so researchers design trials and sample patient populations with the goal of demonstrating the product's effectiveness and safety (Sidebar 1). As we will see, there are many choices to be made in this process (patient characteristics for the trial, sample size, duration of treatment, etc.), all of which force hard choices on time-to-market, cost, risk, and ultimately value.

Sidebar 1: GSK CEO on Cost of Failure

"If you stop failing so often you massively reduce the cost of drug development… it's why we are beginning to be able to price lower. It's entirely achievable that we can improve the efficiency of the industry and pass that forward in terms of reduced prices."—Andrew Witty, Chief Executive Officer, GSK (At the Healthcare Innovation Expo in London, March 14th, 2013 in a speech entitled: Innovation through Collaboration; Leveraging the Expertise of Industry)

Once allocated, R & D funds are tied up for, on average, 2–3 years[1] per phase with little to no visibility into the productivity of the investments made and little to no ability to reallocate these funds to more productive opportunities. Of the funds allocated through this process, the chance that the company loses its entire investment (i.e., the trial is a failure) ranges from approximately 30–55 %, depending on the phase of development.[2] This is not, however, the end of costs to the organization. Typically, as a result of failure, further expenditures are required simply to unwind research programs.

Because of the tremendous uncertainty associated with trying to demonstrate a product's safety and efficacy in real-world trials, companies typically seek to derive value from R & D programs by designing increasingly larger, de-risking trials from Phase 1 to Phase 2. This de-risking process, however, can destroy value in other ways. To increase confidence in the asset before committing significant capital to a long and costly trial, companies typically run multiple earlier studies which extends the overall development timeline, increases costs and reduces the net value of the program. Other decisions that can significantly impact a program's overall value include the fixed costs of trial initiation, per-patient costs, and internal R & D management costs as well as the significant time companies dedicate to analysis between development phases—often 6–12 months or more—which can further reduce the program's value by delaying time-to-market (Table 3.1).

In addition to the inherent complexities and pitfalls of drug development, inflexible corporate processes and infrastructure further hinder pharma companies' ability to deliver value from their R & D pipelines. Rigid budgeting processes focused on financial metrics like EPS lead to a situation in which the finance department essentially dictates a budget to R & D based on strategic priorities and what the organization believes it can afford. This is done without a real understanding of what R & D budget or allocation across R & D programs will ultimately create the most value for the company. The R & D team then takes that budget and distributes it across various drug development programs, after first subtracting fixed costs

[1] Based on average length of development for Phase 2 and 3 trials; see Table 3.1 for additional detail.

[2] Based on average probability of success for compounds in clinical development; see Fig. 3.2 for additional detail.

Table 3.1 Average clinical trial costs and duration by phase

Phase	Costs ($M)[a,b]	Duration (years)[c]
Phase I	$4,211	1.5
Phase II	$6,096	2.5
Phase III	$17,392	2.5
Approval	$4,033	1.5

[a]*Note*: all figures include company-financed R & D only. Total values may be affected by rounding
[b]Pharmaceutical Research and Manufacturers of America, PhRMA annual membership survey, 2013
[c]Paul et al., "How to improve R&D productivity: the pharmaceutical industry's grand challenge," Nature Reviews Drug Discovery, Volume 9, March 2010

(FTEs, etc.) and funding for ongoing trials. The new investments are then "fixed," without real visibility into their productivity and with very limited ability to reallocate resources as the programs evolve—which further increases the overall risk to the portfolio and reduces the company's ability to derive value from it.

These factors all contribute to the challenge of delivering value from pharma R & D. Because clinical development is risky, R & D teams seek to increase the odds of success with large, expensive trials and costly oversight processes—which, paradoxically, can actually leach value from the program instead of increasing it. At the same time, finance teams see the low ability of R & D to deliver value and try to limit their investment, rather than manage and optimize it. Left with limited resources, R & D teams must then choose a smaller number of projects to fund, and then try to increase these programs' odds of success by spending even more on them to increase their sample sizes—and thus the pharma value destruction cycle continues.

Why this Matters: Less Funding, Fewer Novel Drugs

As the pharmaceutical industry struggles to deliver value, investors have slowly turned away from the industry leading to fewer novel drugs coming to market. CFOs, investors, and lenders are becoming increasingly uneasy about investing in this industry and have become more focused on understanding and managing the return on their investments. One can see this in the funding patterns for early stage projects (Fig. 3.3) and in C-suite and board-level discussions centered on R & D productivity.

These funding challenges come on top of an increasingly difficult environment for innovation. Much has been said about the challenges of identifying new drug candidates, as the "low-hanging fruit" is largely considered to be gone. Furthermore, the FDA has indicated its focus on safety issues, particularly in chronic indications like diabetes and lipid control with large patient populations that would otherwise make novel drugs serving these patients particularly valuable. These factors

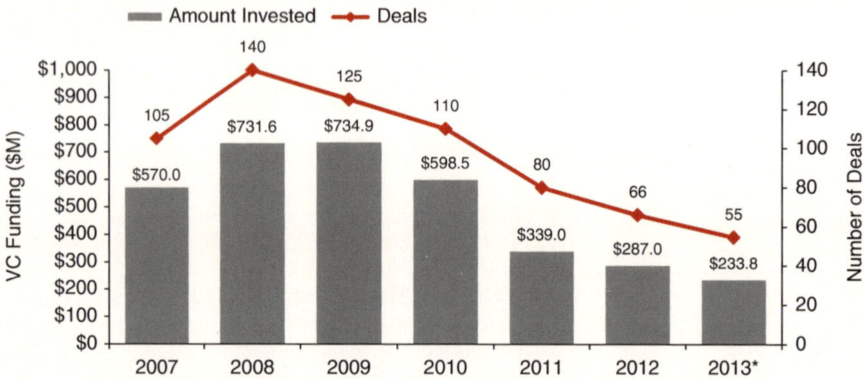

Fig. 3.3 VC funding of seed/start-up biotechnology (2007–2013). PricewaterhouseCoopers and the National Venture Capital Association's MoneyTree™ Report

compound the costs and risks of developing novel drugs and, in many circumstances, decrease the revenue potential of end-markets, further pressuring the value equation for innovator companies.

As a result of these trends, fewer truly novel therapies are coming to market. While the number of FDA approvals of NMEs/NCE in 2012 outpaced that of other years, the number of truly novel therapies is not keeping pace. In place of truly novel science, many companies are pursuing "me too" therapies that are often less risky development projects that can come to market more quickly. This strategy, however, is also under significant pressure as payers begin to demand data that clearly demonstrates the cost-benefit of the new drugs. This phenomenon has been observed in Europe for years where one-payer systems are governed by pharmacoeconomic bodies like NICE in the UK or IQWiG in Germany. With yet one more avenue to value creation being closed off, it remains unclear as to how pharma companies can change the value equation in their favor. Breakthrough science alone is unlikely to be the answer and cost-cutting measures are only a short-term fix. The more likely path is to reengineer the manner in which new science is pursued and demonstrated in clinical trials—a path that can lower risk, speed products to market, lower costs and help companies demonstrate the "value" of their products to payers.

What to Do About It: Maximize Value by Introducing Flexibility/Optionality into Trial Design, Portfolio Management, and Pipeline Funding

Clearly as this cycle of pressured valuations and shrinking funding continues, something must be done in order to ensure continued investment in much-needed innovation to improve patient care. Given the complexity of these challenges, however, pharma companies need to look critically at what they can do to achieve this. We

Table 3.2 Potential adaptive trial designs

Trial adaptation	Description
Phase 2/3 hybrid	• Phase 2 and 3 are run seamlessly as one continuous trial using a combined analysis of all Phase 2 and Phase 3 patients
Interim analysis	• Enables a trial to be stopped early based on futility or efficacy
Sample size reestimation	• Sample size can be increased following interim analysis to achieve predetermined conditional power
Patient enrichment	• Following interim analysis, patient population may or may not be enriched for a particular characteristic or feature
Dose finding	• Multiple doses are tested and patients are switched to the most promising doses over the course of the trial

believe the answer lies first in rethinking the way R & D teams pursue the science and secondly in reassessing how these efforts are financed.

When developing a strategy to "pursue the science" (i.e., the design of a trial or program of trials), companies need to more closely link the design to the *value* of the asset—the discounted expected free cash flows. As indicated earlier, this can be achieved by pursuing trials that focus on lowering risk, lowering costs, and shortening the development timeline. While designing around these imperatives will not be new to R & D executives, a systematic approach that looks at how different designs directly influence value—and what about each design improves or erodes value—is likely to be new. Implementing this process requires detailed discussions between a very diverse set of professionals across the organization: clinical trial managers, bio-statisticians, finance, etc.

While multiple approaches can help companies achieve these "better, faster, cheaper" designs, adaptive trial designs are particularly well-suited to the challenge. Adaptive trials improve on traditional designs by using the information that emerges from ongoing trials to help the management team make more real-time decisions based on the data coming from the trial. A variety of different FDA-endorsed adaptations are already in use today that can increase the probability of success (i.e., sample size reestimation), create the possibility of earlier failure or success signals (i.e., interim analyses), and shorten the overall development time (i.e., Phase 2/3 hybrid designs) (Table 3.2). What this can ultimately mean for companies is accelerated timing (e.g., earlier launch) and/or reduced trial costs (if the number of patients enrolled is decreased). Each of these enhancements will have a direct impact on the value of the trial.

Not all clinical trial programs, however, may be conducive to adaptive trials depending on the natural history of the disease of interest, established endpoints and organizational considerations. In these cases, there are also ways to try and improve upon the financial value of the trial without leveraging adaptive designs. For example, companies could instead focus on reducing the amount of time spent analyzing the data between Phase 2 and 3 or doing sensitivity analyses around probability of success, rather than an unknown effect size.

In addition to improving the financial value of trials, adaptive designs can also be beneficial for patients. For example, take the case of a drug that has no actual therapeutic benefit being administered to patients with life-threatening morbidities (e.g., cancer, infections) in a clinical trial. Obviously, one would want to know this as quickly as possible. Yet, on average 40 % of compounds in Phase 3 trials fail to demonstrate the anticipated therapeutic benefit, putting thousands of patients at risk.[3] An earlier read on a compound's safety and effectiveness would benefit not only drug companies but also the thousands of patients enrolled in trials every year.

Finally, rethinking the design of trials can help to isolate some of the many risks in trials that currently are all grouped together. For example, an interim readout can help to isolate failure risks early in the trial. When an interim analysis with hard continue/stop parameters is incorporated in a trial, it creates a natural segmentation of risk into early, more risky investments (i.e., the first part of the trial where no new information is available to investors) and later, less risky investments (i.e., the subsequent part of the trial once initial real-world data is observed and indicates that the drug is on-track to demonstrate safety and effectiveness). These two different risk profiles are likely to appeal to different kinds of investors. A venture capitalist, for example, might be interested in funding the earlier part of the trial (high risk, high reward), while a royalty investor might be interested in funding the second part of the trial. Indeed, this has been seen in the case of Sunesis, whose drug vosaroxin was funded by Drug Royalty Corp only after the interim analysis of the data had de-risked the investment.[4]

Slicing up the risk in a development-stage program can help improve value by enabling investors with a certain risk–reward appetite to invest only in risks that appeal to them—and avoid risks that do not fit their knowledge or investment profile. This allocation of risk to the most appropriate buyers can help to reduce the overall cost of capital to companies as investors are likely to be willing to offer more fair value for the risks they are investing in if they are not comingled with risks that they do not want to invest in. This separation and allocation of risks to different kinds of investors is seen in the life cycle of many emerging biotech companies that tend to draw funding from increasingly risk-averse investors as the company matures and de-risks its investments. Inside a large, integrated pharmaceutical company, however, this is much more challenging. Redesigning trials to segment the risk is one way to ensure full value is realized by more efficiently pricing the risks of a given program or portfolio.

Companies can further manage and optimize the value equation by developing a more real-time capital allocation program that regularly reassesses where the company's capital will be most productive and generate the most value. In practice what this would look life if fully implemented would be a "financially adaptive" pipeline where the funding of each trial in the portfolio "adapts" in real-time to information coming from ongoing trials. Adaptive trial designs enable this by bringing in more

---

[3] DiMasi et al., 2010.

[4] Sunesis company press release, "Sunesis Pharmaceuticals Announces Closing of $40 Million in Previously Announced Royalty and Debt Financings," September, 20, 2012.

real-time information and allowing for flexible investment points. For example, adjusting sample size in future trials based on information from ongoing or recently concluded trials could de-risk future trials and make the portfolio more financially adaptive. Looking at this at the portfolio level is particularly important since failure in one trial increases the "risk" that the overall portfolio will not meet its required financial return objectives. Therefore, "de-risking" the remaining programs in the portfolio is required in order to maintain the overall risk-reward objectives of the portfolio. Successful implementation of value-based trial designs and real-time capital allocation should ultimately open the door for more funding and, potentially, cheaper funding of R & D.

The last and critical step to this process is communicating the enhanced financial profile (i.e., value) of the pipeline to relevant internal and external investors (e.g., public investors, partners, creditors). By explicitly linking each trial's design and the flexible allocation of capital across the portfolio to value, management should be able to articulate much more clearly exactly how they will bring products to market faster, with fewer expensive failures; how the portfolio is being managed for value; and how all of the changes to trial design are driving greater value. In an efficient market, this communication would help to lower the overall cost of capital for the company as investors begin to understand and price in the reduced risk and increased liquidity of the company's investments in its pipeline. We believe these changes are a key element to helping bio-pharmaceutical companies reignite value generation and break the cycle of low valuation and limited funding.

What this Would Look Like: Big Pharma Case Study

To explore a real-world opportunity to link trial design and valuation, we analyzed a hypothetical new agent in acute myelogenous leukemia (AML). This example embodies several key challenges in pharma R & D. First, recent advances in genetics and preclinical biology have led to the identification of many new targets and asset classes in previously under-studied diseases, such as AML, in which there are often insufficient data to refine or de-risk the choice of indication or patient subset. Second, pharma R & D groups increasingly struggle with how to develop assets with huge unmet needs and scientifically attractive hypotheses but small commercial opportunities, such as in AML. Third, in many indications (including AML) there are no well-vetted alternatives to the registration endpoint (in this case, overall survival (OS)), making it hard to de-risk the Phase 3 study with a shorter, cheaper Phase 2 trial. Thus, like many current R & D opportunities, new agents in AML appear on their face to be scientifically and clinically attractive, but high-risk with a low projected ROI.

Our base case development plan in AML was, as we predicted, unattractive from a value standpoint (Fig. 3.4). Because there is little evidence to support any other endpoint besides OS, the Phase 2 trial is essentially a smaller version of the Phase 3 study. The average time (68 months), cost ($35.7 M), and estimated POS (59 %) of

Average Trial Parameters for Base Case and Adaptive Scenarios*

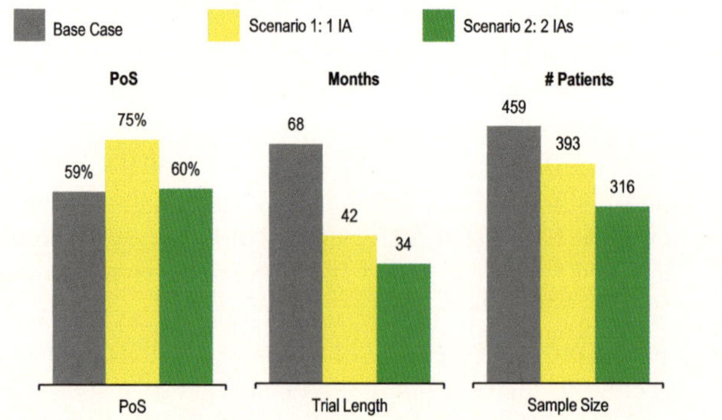

Pre-market R&D Costs *(per patient)*	
Phase 2	$60,000
Phase 3 and Adaptive Trials	$90,000
Additional Costs for Adaptive Trials	
$275,000 / trial	

Valuation Assumptions	
Discount Rate	15%
Tax Rate	35%
Valuation Date	12/31/2012

Approach

- Monthly development costs and sales are modeled for each possible option within each scenario
  - E.g., succeed early, fail early, succeed at required # of events, fail at required # of events, etc.
- Pre-market R&D costs modeled on a monthly basis
- Per patient costs are representative numbers and assumed to be distributed over the course of the treatment duration with costs beginning to accrue as each patient is recruited
- Costs and sales estimates are then probability adjusted based on Cytel analysis
- *These estimates are then rolled up for each scenario and feed into overall program valuation*

*Note: Trial length and sample size for base case and scenarios based on average levels observed across >10,000 simulations
Base case design includes probability of having 9 month gap between Phase 2 and 3, if the former is successful
Source: Cytel analysis

Fig. 3.4 Representative metrics for base case and adaptive designs

the program are not unreasonable based on industry standards. But because AML is a small indication with an established standard of care, this base case plan yields an extremely modest eIRR (3.1 %), well below the observed industry average of ~8 % (in the 2010–2013 timeframe).[5] Thus, the value of the program is limited by the long, expensive, and risky development plan, despite its potential to be a valuable new agent to advance clinical care.

Given the limitations imposed on this program by traditional trial design, we explored how we could enhance the value of this program with different approaches to trial design. To do this, we developed two scenarios with different objectives that each incorporated several adaptations to decrease time, cost, and/or risk compared with the base case. The primary objective of the first scenario, "Scenario 1," was to decrease the time to the first data readout relative to the base case. The second scenario, "Scenario 2" was designed primarily to maximize the value of the program, regardless of when the first data readout occurs (Fig. 3.5). Importantly, both of these designs benefit from the fact that adaptive designs do not require waiting for data maturation after the first stage. We review the approach and outcomes for each of the scenarios in more detail.

[5] "Measuring the Return from Pharmaceutical Innovation: 2013," Deloitte Centre for Health Solutions and Thomson Reuters.

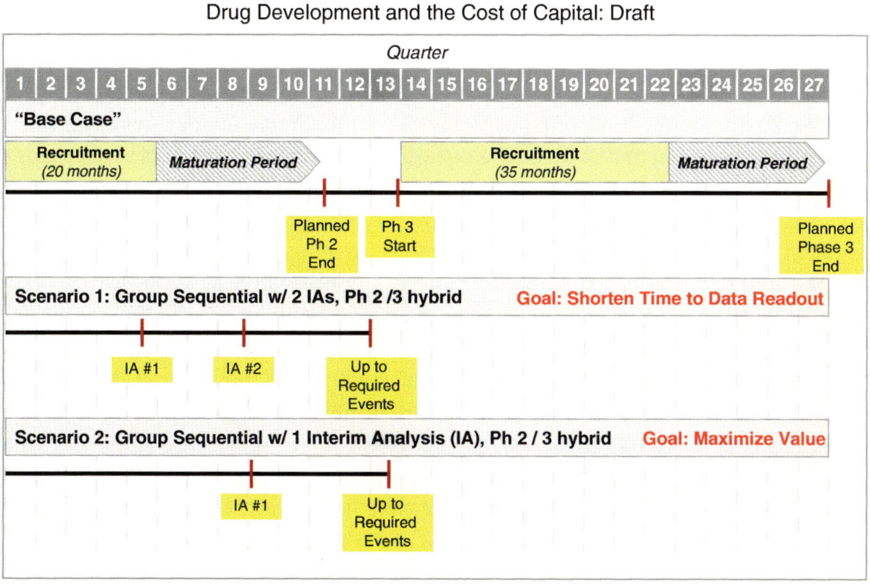

Fig. 3.5 Trial designs (base case and adaptive scenarios)

Scenario 1

In Scenario 1, we used a combined Phase 2/3 trial and two interim analyses. With the combined Phase 2/3 trial, we eliminated the duplicated effort in the sequential design and reduced the overall clinical development duration. The two interim analyses enabled better alignment of cost and risk throughout the program.

Scenario 1's adaptive design leads to a significant increase in the program's risk-adjusted NPV, from \$5.1 M to \$34.9 M. This is primarily driven by the dramatic 55 % shortening of the total development time, which both accelerates the launch and adds to the total time that peak revenues are realized before loss of exclusivity (LOE). Notably, because the accelerated launch pulls forward all of the revenues (raising the present value of on-market revenues), this factor alone contributes almost half of the additional program value generated from the new design. Combining the two studies into a single trial also increases NPV by reducing the total program cost (via decreased sample size), even taking into account the additional administrative expense of executing an adaptive trial. Although the new design does lead to somewhat higher "up-front" costs, this is a minor increase (\$10.5 M in the base case vs. \$18.1 M for the adaptive design in the first 2 years of the program). Overall pre-market R & D costs, however, ultimately end up being lower for the adaptive design (\$35.7 M for the base case versus \$26.3 M for the adaptive design).

Scenario 2

In Scenario 2, we leveraged a combined Phase 2/3 trial and one interim analysis. Similar to Scenario 1, the combined Phase 2/3 trial eliminated the duplicated effort in traditional design and reduced the overall clinical development duration, though to a lesser degree given the differing objective of this design, The use and timing of the interim analysis drove not only a decrease in clinical development time (45 % reduction) but also a significant increase in probability of success (25 % increase from 59 to 75 %). The probability of success improves because the patients that normally would have been studied in Phase 2 (and not counted in the sample size for Phase 3) are now included in the final Phase 3 readout, effectively increasing the power of the important, late-stage trial. Additionally, it was possible to decrease the sample size by ~15 %. Ultimately, as a result of these changes, this design resulted in an increase in NPV from \$5.1 M to \$42.3 M.

Besides enabling a feasible and attractive path forward in AML from a clinical and financial standpoint, this analysis more generally demonstrates the powerful link between R & D decisions and financial value. Although many R & D executives understand the need for rapid R & D, this example shows the actual financial value of decreasing development time—and therefore helps enable them to determine the right amount of resources to spend in exchange for speed. From this standpoint, adaptive trials, which are usually significantly faster than their non-adaptive counterparts, provide significant benefits that have been incompletely leveraged by large pharma organizations.

This example further shows that independent of trial design considerations, R & D cost containment may not be nearly as critical a driver of a program's financial value as increasing program speed. For example, reducing the time spent in review and decision-making between program phases is likely of far greater value than incrementally reducing per-patient costs. As demonstrated here, the largest determinant of "R & D cost" is not cost at all, but time—and this is the aspect of clinical development that deserves the most focused attention from R & D executives.

We also show here the need to focus R & D decisions on overall program value, not just on acute budgetary concerns or statistical parameters. In this example, the marginal increase in near-term cost of a Phase 2/3 study over a conventional Phase 2 trial is more than offset by the impact on eNPV and the program's probability of success. This analysis can help align R & D and financial decision makers on the common goal of developing new drugs with both clinical and financial value.

From an R & D portfolio perspective, this analysis has two further implications that could improve decision making, overall R & D productivity and financial returns. First, across the industry there are abundant promising programs that are equally risky without additional clinical data. Currently, R & D executives are often forced to "pick the winners" in this information-poor setting based on a combination of clinical promise and near-term budget availability. This decision making process limits the number of otherwise promising programs that get viable "shots on goal," and also can lead to continued funding even in the face of mediocre clinical data due to the "endowment effect." By linking trial design, investment and value, R & D organizations can test more agents in trials and either scale up the investment or "kill the losers" based on bona fide clinical data.

Finally, transparently linking financial metrics to clinical development decisions and building optionality into the portfolio allows managers to make more rapid, fact-based decisions about where and how to deploy scarce R & D capital to optimize returns: among therapeutic areas and asset classes, over time. Such a cross-portfolio analysis could provide a more rational, data-driven basis for flexibly allocating capital to the highest value opportunities, be they individual assets or entire R & D divisions.

Chapter 4
Investment Considerations for Pharmaceutical Product Portfolios

Raymond A. Huml

Background and Terminology

Pharmaceutical executives with varying backgrounds and expertise are increasingly being asked to provide input on complex strategies related to portfolios of products (which products to keep and which products to advance), including regulatory and commercial competitive intelligence (e.g., is a product worth advancing given the current or future market environment?). Given the appropriate expertise—preclinical, clinical pharmacology, clinical, regulatory affairs professional, CMC/supply chain, commercial, payer, patent law, etc.—they may even be asked to provide input to help estimate the probabilities of success related to key milestones in clinical drug development, such as the likelihood of the product being approved in a particular market of interest.

In an era where product partnering is becoming more common, pharmaceutical executives are also being asked to comment on matters related to product development even earlier in the drug development cycle. For example, it is becoming more common to take account of key commercial insights, payer perspectives and even quality-of-life (QoL) issues into the earliest phases of clinical drug development, such as Phase II. This input can positively or negatively impact business decisions, which in turn, impact the bottom line.

As in medicine, there is an art, as well as a science, to the due diligence activity of estimating the probabilities of technical success for a particular compound or portfolio of compounds. And it is usually not sufficient (from an upper management perspective) to just use a published benchmark to estimate the probability of technical success—such as Tufts University's metrics published in Parexel's Statistical Sourcebook. Therefore, it behooves each pharmaceutical executive to learn more

R.A. Huml, M.S., D.V.M., R.A.C. (✉)
Quintiles Transnational Corp., 4820 Emperor Boulevard, Durham, NC 27703, USA
e-mail: Raymond.huml@quintiles.com

Z. Antonijevic (ed.), *Optimization of Pharmaceutical R&D Programs and Portfolios: Design and Investment Strategy*, DOI 10.1007/978-3-319-09075-7_4

about how regulatory and commercial outcomes impact business decisions. The approval criteria for a recently registered compound in a similar class, for example, provide insights into primary endpoints upon which registration will be based, as well as the mechanism of action. If the product is marketed, sales data from sources such as IMS Health—in conjunction with the right surrogate (e.g., a product analogue used as a benchmark for the product under investigation, which may or may not be in the same product class but should have similar attributes—can even be used to estimate how well a product might fare in the market or within a portfolio).

There is a need to leverage both technical expertise (R & D and commercial) and regulatory experience to improve decision-making by better understanding the rationale behind making complex business decisions that can impact whether products will be advanced, sold, out-licensed, or abandoned.

This chapter discusses some of the fundamental investment decisions based on the author's experience in investing in portfolios of pharmaceutical products—such as how much one can invest (and thus potentially lose) and the palatability of the proposed return structure. The chapter breaks down many facets of this complex process to give the reader insights into business decisions related to portfolios, with the most common situations highlighted.

This chapter also provides several case examples of these principles in action from the author's experience while serving as a member of the due diligence group at Quintiles Transnational Corp. The case studies were created using only publicly available information.

This chapter explores definitions and boundaries of many factors influencing investment success associated with product-based portfolios, whether the definition entails the number of assets available for partnership (whether by choice, inheritance, business model, or available cash) or investment appetite.

Next, there is a discussion of boundaries set up by the capital limit of a fund or artificially created boundary (e.g., special purpose entity [SPE]), and whether the portfolio is a group of assets or the goal is to partner on multiple single assets to have a portfolio effect. The author then discusses the feasibility of capital syndication to garner financing for portfolios, and discusses the key differences in portfolio strategy between big pharma and biotechnology companies.

Finally, the chapter concludes with a discussion of the key differences in portfolio strategy between big pharma and biotechnology companies and covers portfolio management, innovative partnering solution strategies and commercial perspectives.

Because some business terms may be unfamiliar to pharmaceutical executives, a list of terms and acronyms is provided in Table 4.1.

Introduction

Investing in the pharmaceutical industry is a risky business. It behooves each person involved in the investment process to understand the proposed transaction well as the processes used to mitigate any risks that are identified during due diligence.

Table 4.1 Business terms and acronyms

Term or acronym	Definition/explanation
NPV	Net present value is a standard method for estimating the present value of an asset taking into account the (ever reducing) time value of money over the life of a proposed deal (e.g., 5 years). NPV compares the present value of money today to the predicted value of money in the future, taking inflation and returns into account.
Risk-adjusted (RA) NPV	The risk-adjusted (RA) NPV is adjusted down based on the fact that it is not 100 % assured that any drug will get approved by any regulatory authority. In its simplest form, RA NPV is a numerical value created from the NPV multiplied by the probability that the product will be approved (e.g., \$100 m × (PoS) 80 % = \$80 m). In reality, there might be multiple factors that could drive down the PoS for a given product, including, for example, an unfavorable safety characteristic, a CMC or supply chain uncertainty, or the potential for a lawsuit against the product's manufacturer.
PoS	Probability of success is the probability that a milestone will occur. Often referred to as likelihood of regulatory registration, this can apply to any commonly known or artificially created endpoint (end of Phase II, primary efficacy endpoint, successful filing, last patient successfully enrolled in a trial, a certain amount of weight loss (in pounds)/unit of time, etc.).
Third-party capital investor	A third party capital investor, such as a venture capital investor, can provide a source of capital in addition to the buyer or the seller in a product-based transaction.
Return structure	The return structure defines how the capital provider will be paid back. It may involve milestones, royalties, equity, services or a combination of different mechanisms.
Venture Capital (VC)	Venture capital is financial capital provided to early-stage, high-potential, high-risk, growth startup companies. The typical investment occurs after the seed funding round and functions as a growth funding round (also referred to as Series A round) in the interest of generating a return through an eventual realization event, such as an Initial Public Offering (IPO) or trade sale of the company. VC is a subset of private equity.
Concentration limit	For many providers of capital that garner capital from multiple sources (e.g., "funds"), there is a predefined limit to the maximum amount of capital that can be employed for any one deal. This is related to the magnitude of the fund, with higher limits for funds with greater amounts of capital.
Monetization	Monetization is the process of converting or establishing something into legal tender (cash). For product based investing, it is way to obtain value in the present for perceived returns that were originally contracted to occur over a period of time in the future. For example, if a return of 20 % was reasonably expected at the end of a 5-year period, the expected return could be sold for a lump sum prior to the end of the 5 years, usually for a lower price, such that the profit (and risk and potential upside) would be transferred so that immediate cash could be secured.
Private equity firms	In the past, private equity firms organized limited partnerships to hold investments, in which the investment professionals served as general partners and the investors, who were passive limited partners, put up the capital. A compensation structure, still in use today, also emerged with limited partners paying an annual management fee of 1.0–2.5 % and a carried interest typically representing up to 20 % of the profits of the partnership.
Buyer	Refers to the entity providing services or cash.
Seller	Refers to the entity holding the intellectual property of the asset/product in a particular geography.

Initial investment terms are proposed in the first iteration of a "term sheet." The term sheet is usually a moving target and thus, subject to change and negotiation, until terms are finalized in the contract.

Investments can take many forms, from purchasing stock (or entire companies) through investing in individual products (in development or commercialized). This chapter focuses on the latter investment option.

Investment groups employ a myriad of strategies to dilute risk. For example, to mitigate the risk associated with a single product asset, an investor group may consider accumulating a portfolio (a *portfolio* is defined as at least, but typically more than, three products). A *product*, for purposes of this chapter, is defined as a single small molecule, biologic or medical device.

While each individual product may have a high potential for technical failure, a portfolio of at least three compounds will increase the chance for at least some success. The logic is that, with a portfolio of products, there is a greater likelihood that capital will be returned from the success of at least one predefined milestone, even if others do not meet certain milestones, such as proof of concept (PoC), a primary efficacy (or another predefined clinical) endpoint, progression to the next phase of drug development, regulatory approval or a certain sales forecast. A commonly used expression describing this mitigation strategy is "having more shots on goal."

Most pharmaceutical companies acquire or develop products with the intent of forming a portfolio. Decisions are based on the company's business model, resources and capabilities. For example, Genzyme, prior to its acquisition by Sanofi, focused almost exclusively on niche, orphan drug products (that allowed for expedited regulatory review and exclusivity advantages) with very high costing models. For example, although their orphan drug-designated products were used in only a small number of patients, the cost/treatment was very high (e.g., >$100,000 USD per year).

This chapter will not discuss the many decisions regarding acquiring and advancing candidates in a portfolio (usually based on net present value [NPV] driven by commercial considerations), but rather, on the processes used by third-party capital providers to select investment candidates for a portfolio of products.

The two main drivers for consideration by any third party when investing capital in a portfolio are: (1) the magnitude of risk that the investor (e.g., third party capital provider) is asked to take; and (2) the palatability of the return structure to the seller. Most deals, in the author's experience, are abandoned or grow cold not because of a flaw in the product or company but because of a factor related to the proposed return structure.

Controlling Risk

Risk for the investor is scaled, with the greatest exposure associated with earlier stages of drug development (e.g., preclinical or Phase I). Because of the high clinical uncertainty with products in preclinical development, the author has seen some investment companies value these assets as being worth very little or even nothing until a milestone, such as IND readiness or submission, has been achieved.

With a few exceptions (e.g., accelerated approval in high-risk settings such as oncology), risk lessens as the product is studied in greater numbers of patients and in later stages of development (e.g., Phase III or even post approval, such as estimating the chance of line extensions for approved products).

Risk also declines due to the portfolio effect, where the riskiness of the total portfolio decreases as more assets are added. In addition, some companies have used other strategies that employ the portfolio effect; for example, creating a portfolio with a sufficient number of assets based strictly on drug development averages (usually either calculated by a due diligence team or using published averages) as a way to assess the likelihood of hitting a milestone with a basket of early-stage (e.g., Phase I) compounds. In this type of scenario, a portfolio is created by adding together three or more compounds. As a theoretical example, if the average likelihood that one compound (in a certain therapeutic area) will meet a predefined milestone is 10 %, then, by having at least ten compounds in the portfolio (without "cherry picking" or biasing asset selection criteria), it should be $1-0.9^{10}$ (about 65 %) assurance that at least one of the ten compounds will meet the predefined criteria.

To further reduce the risk associated with portfolios, companies that provide capital perform due diligence to better understand that risk, with the goal of protecting themselves via contractual language. Due diligence includes a thorough investigation of all available data—both proprietary (after a confidential disclosure agreement [CDA] is signed/executed) and nonproprietary—by a team of experts with the intention of predicting future events (such as the probability of success and the sales forecast). This process aims to allow a deal team to manage risk and execute a favorable financial transaction. All venture capitalists and investment arms of companies perform due diligence to a greater or lesser degree prior to investing money in a deal or getting involved in a partnership arrangement.

The due diligence exercise, culminating with a report that articulates the team's findings, should inform capital providers of the positive, negative and unknown attributes of each drug/biologic/device in a portfolio (or proposed portfolio) and the probability of successful achievement of a predefined milestone. While this milestone is typically regulatory approval, milestones can also be defined as meeting the primary efficacy endpoint, advancing to the next stage of drug development or even meeting an artificially created endpoint (e.g., decrease in HbA1c, % weight loss, % decrease in BP, etc.). Due diligence also can provide an understanding of the sales potential of the product for deals where returns are matched with royalties or where a risk-adjusted NPV of an asset is required.

Return Structure

The due diligence exercise is aligned with the proposed return structure. Therefore, only the findings that could potentially affect the return should be reviewed. It is not important to conduct a full commercial due diligence exercise, for example, on a portfolio of Phase II assets whose repayment is based entirely on milestones.

Return structures can be aligned and negotiated based on the perceived risk. For example, if the product is a "me-too" with a good chance of regulatory approval, the buyer may stake the return to that milestone, whereas the seller may wish to defer a return to a royalty component. This may be less certain (and more risky for the capital provider), given the strong competition within a class of therapies where there are alternative options for patients.

Another return structure could include a mechanism to decrease the cost of capital (CoC). For example, a common method is using a seller's existing royalty strip to offset the CoC. A *royalty strip*, for purposes of this chapter, refers to an ongoing (usually monthly or quarterly) payment made to a seller from the sales of an approved (and different) product in return for an investment made in an earlier transaction. This existing and expected continued flow of capital can be transferred to the buyer. Royalty strips are usually only a fraction (e.g., <10 %, thus the term *strip*) of the total sales of a particular product and may be confined to a particular geography (e.g., USA only, EU and USA, Canada only, etc.).

Number of Assets Available for Partnership

There is no ideal number of assets to be included in a portfolio. The number is limited by the amount of capital available for such a portfolio (via a single source or a syndicate). The optimal choice of assets for a portfolio can be aided by estimating the risk-adjusted net present value (RA NPV) of a set of possible asset choices. In addition to historic data about the potential asset, the estimation of RA NPV usually needs to include prior beliefs in the quantification of risk pertaining to the asset. Bayesian statistics are appropriate and naturally incorporate prior beliefs, and may be used here. Using such techniques, the optimum number of products in a portfolio can be aligned with the risk associated with each asset, with the differing probabilities of success (for predefined milestones) aligned such that given a range of prior beliefs, the capital provider can minimize the chance of loss stemming from the investment. In the author's experience, the number of assets usually ranges from four to eight, but could be higher or lower, and is also based on investment appetite (discussed below). Additional discussion of Bayesian statistics is beyond the scope of this chapter.

In the author's experience, many larger companies, mainly through mergers and acquisitions, are sitting on large portfolios of less-strategic clinical assets (e.g., assets that lack allocated budgets or have been de-prioritized for a variety of reasons). They may wish to partner on these assets (to retain the value inherent in the intellectual property should the compound be approved or to increase the investment value of each candidate in later stages of development) for the purpose of outsourcing the asset or taking it to the next stage of development. In general, investor groups wish to partner with companies whose available assets are of high strategic value. Lesser-value products may not get the capital or attention of the partner needed to optimize their success.

Some companies want the potential capital partner to take all of the risk for a very large, but typically very unlikely, upside. This scenario, however, does not generally align interests.

To reduce the risk for each stakeholder (especially the third-party capital provider), it is advisable for the buyer and the seller to invest in a substantial portion (e.g., as much as half) of a program/portfolio. While this may not be an option for certain potential investment opportunities, it helps aligns partners and ensures that each partner is doing all that it can to optimize the relationship. This investment alignment paradigm is commonly referred to as "putting skin into the game."

Larger portfolios (e.g., with more than three assets) are usually limited by the amount of capital required and degree of risk. For example, a basket of Phase I assets would carry very high risk but lower costs, as clinical drug development costs in this phase are much less for each trial compared with those in the later stages of development (e.g., Phase II or Phase III). A basket of Phase III trials, on the other hand, given the clinical and efficacy data accumulated to date, would be considered much less risky. However, the typical cost of these pivotal trials would probably make it difficult to attract capital from a single source for more than say, three studies.

Risk may also depend upon whether the drug candidate is in a class of compounds with proven efficacy in the disease area being tested or is based on an unproven scientific theory. For example, many new compounds being tested in oncology, Alzheimer's disease and other areas of unmet need have inherently more risks because the cause or the pathway of the disease is not yet fully understood. In contrast, a compound with the same mode of action and therapeutic class as an approved drug would be expected to have a much higher probability of success.

Investment Appetite

Investment appetite is correlated with the degree of capital loss that an investor group is willing to risk. No investor group wishes to lose money, but due to the business plan or magnitude of capital available for investment (and thus the ability to diversify), some investors may be willing to take less upside in exchange for the ability to do more deals, while others may want to take a larger upside to satisfy their shareholders and investment committees. In general, the greater the risk the capital provider is asked to take, the larger the expected return. This general paradigm provides the basis for negotiations from the initial term sheet to the final contracting stage of the partnership.

As noted earlier, many deals are not executed for reasons other than a flaw in the product or the company. In addition to a variance in investment appetite, reasons may include ongoing or perceived patent issues, inadequate (or unobtainable) contractual protection, inability to obtain a warrant or representation (e.g., legal protection) or the desired accounting treatment (Generally Accepted Accounting Principles or GAAP in the USA vs. International Financial Reporting Standards or IFRS in Europe), an unsuitable return structure (e.g., not enough return for the perceived risk associated with the product), a future risk that is deemed unacceptable (e.g., a potential lawsuit), or an unreasonable return structure (as deemed by the potential partner).

Capital Boundaries

A *special purpose entity* (SPE), for purposes of this chapter, is defined as a legal entity (usually a limited company of some type) created to fulfill narrow, specific or temporary objectives. SPEs are typically used by companies to isolate the firm from financial risk. It was noted earlier that portfolios can be limited by the capital limit of the fund. Portfolios can also be limited by the boundaries created by syndication in an artificial environment, such as the SPE.

Most investment groups, due to predefined investment criteria, have a limit to the amount they can invest in a particular deal and this limit is typically based on a percentage of the entire fund amount. For example, if a company has a $100 million fund and the concentration (maximum) limit is 20 % per deal, it can invest in only five deals that cost no more than $20 million each unless it invests in deals of smaller magnitude or is part of a syndication process.

In the author's experience, many companies realize that they cannot obtain all the capital required for a large portfolio of products from one source because of internally defined boundaries (which may be driven by P & L relief or some other factor). In this case, the magnitude of the limit, for example $250 million, will dictate how many candidates can be accommodated within the boundaries of this limit—e.g., fewer Phase III candidates and more Phase II and Phase I candidates.

For companies that provide pharmaceutical services as a form of capital, such as clinical research organizations (CROs), it is important to note that while the Phase II stage of drug development presents greater risk, there is also more room for influence, in contrast to the Phase III or pivotal stage which offers less chance of influence (for a variety of reasons including regulatory interactions). In general, by the time of Phase III clinical development, the optimum dose has already been identified and the pivotal trial designs have been agreed upon with the regulatory agency (e.g., FDA's End-of-Phase II Meeting).

Portfolio Effect

Some companies have a business plan that entails partnering on a number of single product assets, usually at a later stage, such as Phase III, with the goal of obtaining a greater number of assets to dilute risk and obtain a "virtual" portfolio effect. This can be an effective strategy if it can be accomplished quickly, as the risk is not optimal until the full portfolio effect is achieved.

However, in the author's experience, many third-party capital providers are reluctant to be first in an investment vehicle with single-product assets with the goal of a portfolio effect—especially with portfolios of higher-risk, e.g., a basket of Phase I or Phase II products.

Phase I vs. Phase II vs. Phase III

It has long been known in industry that, by combining a group of products, especially in one therapeutic area, the risk of total capital loss for the group is reduced. In essence, risk is diluted. But there is an art to evaluating the risk vs. cost of groups of products in different stages of clinical development. For example, when looking at a basket of Phase I studies, there is usually insufficient evidence to draw a definitive conclusion on either efficacy or safety. At the other end of the spectrum, for a basket of Phase III products, the cost to run the trials makes such an experiment prohibitive for any one company and it is unlikely, with current competition models, that large pharma companies would be willing to place their key Phase III assets into a separate entity. What about Phase II? Perhaps this might be the place where a balance could be found between cost (lower with Phase II development) and the likelihood of being able to draw a definitive conclusion at the end of each trial.

Syndication

Theoretically, syndication (e.g., for larger portfolios) should be an easy, straightforward way to gain access to additional sources of capital. Syndication, for purposes of this chapter, refers to having more than one capital provider providing capital required for an executed portfolio deal.

In reality, however, syndication is very difficult and often impossible. Leaving aside the financial climate, other reasons can include:

- Timing—the point at which a potential partner is brought to the table for a term sheet (the earliest stage of negotiations between a buyer and a seller whereby the return dynamics and risk assessment evolves as due diligence is undertaken) or contracting (when due diligence concludes, the deal terms are finalized and lawyers draw up a final contract which is signed by both the buyer and the seller upon deal execution) discussions.
- The desire to have disproportionate influence on the deal (e.g., a $20 million capital provider in a $100 million deal wanting undue influence on contractual protections).
- Differences in desired investment returns (one party willing to take less vs. one partner seeking greater returns).
- The novelty of the transaction.
- The ability of the potential partner to raise more funds.
- Perceived differences in the risks estimated by different due diligence teams.

Because of the large number of expert resources required to perform due diligence (functional area experts in preclinical, clinical pharmacology, medical, CMC/supply chain, regulatory affairs, commercial, sales analytics, patent law), plus deal term negotiators, some companies will "piggyback" their investment on top of a credible, additional third-party capital provider with the primary goal of reducing cost to invest. This can be helpful, but then that third-party capital provider may lack control and be at the mercy of the conclusions reached by the other provider's due diligence team.

Key Differences Between Big Pharma and Biotechnology Companies

Based on the author's experience in partnering with biotechnology companies, this section outlines several successful approaches that provide alternatives to classical partnering with large pharma and venture capital entities. These solutions remain beneficial and financially profitable to both parties over the long-term (i.e., 3–5 or more years).

The Productivity Dilemma

Today, the research and development (R & D) pipelines once enjoyed by large pharmaceutical companies are no longer producing a sufficient number of drug candidates. As pipelines shrink, the pharmaceutical industry has tried to develop novel solutions, such as creating "incubators" or "accelerators" where efficiencies in smaller development companies can be leveraged to expand pipelines. Often these smaller biotech and pharma companies are the innovation engines filling major pharmaceutical companies' R & D pipelines. The bigger players continue to invest in the smaller companies in exchange for downstream milestone payments and royalties, or even purchase options on one or several compounds.

Additionally, academic institutions can act as incubators (or sources of new drug candidates): collaboration between pharmaceutical companies and academic institutions provides a mutually beneficial arrangement, as academic centers are not positioned to commercialize a product.

Investments by large pharma can provide the smaller biotechnology company with the capital to build out its infrastructure and further develop its earlier stage pipeline. This type of partnering relationship establishes credibility for the smaller company and creates value for the larger pharmaceutical company without the need to fund and direct its own internal development program. In addition, the ability of smaller companies to adapt nimbly to changes in technology and overcome development hurdles are key advantages that large pharma lacks. Generally, these smaller firms take their compounds to the proof of concept (PoC) stage of drug development and exit to pharma companies that are better positioned for costly late-stage development and commercialization of the product. A key benchmark of this approach is the opportunity to change the accounting dynamics regarding research infrastructure with smaller firms and outsource the burden of fixed cost inherent in development. With this outsourced approach, the smaller company turns the fixed cost into a variable cost to obtain more-desirable accounting treatment.

It should be noted that this risk-shifting approach often carries a high price for the smaller company, as it frequently relinquishes control over the asset(s) and awards a significant share of any potential upside to the large pharmaceutical partner.

However, it is possible to devise partnering alternatives to allow smaller companies to retain strategic and financial control over their assets. Two main goals of such partnerships are to optimize the drug development process by addressing resource constraints and to minimize the risk of innovation by sharing it.

Types of Biotechnology Companies

For purposes of this chapter, three main categories of biotechnology companies are defined based on their corporate maturity. The first is *discovery*, the second is *bridging*, and the third is *standalone*. Each company type has different capital needs related to the number of compounds in the clinic. No biotech firm, regardless of category, is anxious to part with its intellectual property to partner with another entity—especially if the upside is perceived to be significant.

The discovery biotechnology company frequently has a novel approach to identification of drug candidate leads, usually based on a new target or novel delivery system, combination of existing products or novel technology platform, but does not have sufficient infrastructure or cash to advance them all in the clinic. These small entrepreneurial biotechnology entities often live hand-to-mouth existences, working to obtain just enough capital to get to their next R & D milestone. These less-mature enterprises face a shrinking runway and may seek to mitigate risk by partnering with large pharma. Such deals tend to come at a high price. The biotechnology company is often forced into mortgaging a good portion of its future value in exchange for a chance to continue development.

The bridging biotechnology company has advanced compounds into the clinic and has more credibility within the pharmaceutical industry than the discovery company because of existing partnerships or early demonstration of clinical PoC. These companies frequently have partnerships in place with large pharma and seek an alternative and more-innovative financing strategy that will allow them to retain more asset value than in previous partnerships with large pharma.

Standalone biotechnology companies are characterized by sufficient resources from existing revenue streams to advance their own candidates to commercial success. They act like large pharma in the way they deal with other, smaller, biotechnology companies and have their own marketing and sales groups. They tend to either use less-innovative partnering solutions or more-complicated ones, such as those highly correlated with internal accounting standards.

Private vs. Public

Another interesting distinction between biotechnology companies is the different ways in which public versus private companies finance their research and development programs. In general, both private and public companies, especially those not generating earnings, consider their valuations depressed. A brief overview of selected funding option considerations for private and public companies is provided in Table 4.2.

Key Event Drivers

Earnings-driven biotechnology companies are interested in partnerships that allow them to retain control and gain future upside while deferring R & D expenses in the near term. The key needs (or drivers for financing) for biotechnology companies fall into three main categories:

- Financial reporting needs
- Cash needs
- Operational needs (include credible partnering with a company that can deliver services, such as large CROs).

Financial reporting needs are best interpreted in light of acceptable accounting standards (e.g., GAAP in the USA or IFRS in the EU) by qualified parties. In general, earnings-driven companies are interested in mitigating the financial impact of R & D expenses until a compound is marketed. Because accounting factors are subject to expert interpretation, they are beyond the scope of this chapter.

Table 4.2 Selected funding option considerations for private and public companies

Funding option	Private company	Public company
Venture capital	X	
Public (at IPO)	X	X
Institutional		X
Large pharma	X	X
Portfolio	X	X
Royalty	X	X
Monetization	X	X

Note: Both private and public companies will benefit from achievement of milestones (successful clinical results, proof of concept [PoC], regulatory filing, Investigational New Drug Application [IND] submission, etc.) because the achievement of milestones can produce additional boosts in equity. They can both benefit from the announcement of a deal. This news adds credibility to a particular technology or a company's approach to clinical drug development. Such announcements may even boost the stock price for public companies

Cash needs are usually acute. Discovery and bridging companies are often focused on dilutive aspects of financing more than near-term earnings. In general, they also seek to retain control of their assets (e.g., compounds).

Operational needs can only be addressed by credible partners that can provide clinical trial infrastructure or intellectual drug development expertise; therefore, these are limited to major players such as large pharma or global CROs.

Other key event drivers include large milestone payments triggered from relationships with large pharma and initial public offerings, but these events are beyond the scope of this chapter.

Partnering Solutions

Driving demand for alternative partnering solutions is the biotechnology company's desire to retain strategic and financial control over its assets. Two goals for such transactions are optimizing the drug development process by addressing resource constraints and de-risking innovation by sharing risk in clinical and commercial programs.

Portfolios (Phase I or Phase II Funds)

To mitigate the investment risk associated with a single product or earlier stage assets, as mentioned earlier, an investor group may consider accumulating a portfolio of Phase I and/or Phase II products.

This approach can be attractive for companies with platform technologies where many targets can be produced, and, due to cash constraints, only a few can progress to the clinic. Key problems with Phase I portfolios include defining success (e.g., submission of an IND), how to advance the product to Phase II, and how to have enough targets to make the expected risk palatable from a risk–return perspective (e.g., using Tufts University's Center for the Study of Drug Development's published probabilities of success).

Problems with partnering in a Phase II portfolio approach include: (1) Dilution of value by large pharma deals already in place with the biotechnology company; (2) Justifying an investment sufficient to move the product to an acceptable exit strategy based on the risk profile; and (3) Cost of assessment of portfolio risk if not confined to one therapeutic class. The third problem involves increased due diligence costs associated with obtaining extra resources, including additional medical expertise, in order to understand different therapeutic classes in a portfolio of diverse therapeutic products.

For fully mature biotechnology companies, a structured finance deal may provide the greatest flexibility and widen the number of options. *Structured finance*, as used in this chapter, is defined as a risk-based commercialization and/or development arrangement whereby funding and/or services is provided in exchange for fees and/or product royalty rights.

Alliance Management

Pharmaceutical and biotech companies are entering into alliances and joint ventures at an ever-increasing pace to obtain financial and other resources to begin clinical development programs while managing costs and optimizing earnings. Strategic partnering—when done correctly—can offer significant advantages versus "going it alone" in solving these major challenges. Identifying the risks and benefits associated with implementing and managing successful alliances begins with the due diligence assessment by the potential alliance partner.

If they are to make the partnership work, each company must assign a dedicated executive responsible for managing the operations of the product partnering arrangement. This executive typically has the title of alliance manager (AM), and it is the AM's role to champion the success of the alliance and to ensure each partner obtains expected financial returns. This role can also assist greatly in bridging any operational issues with partners that provide services as part of the deal.

In addition to assigning an AM, a governance committee may be formed. The governance committee may sit above other committees, such as a joint operating committee (JOC), and is comprised of upper management executives (with merited voting and non-voting privileges) that meet on a regular basis (e.g., quarterly or semi-annually) to assess progress and make course corrections as necessary. This team is charged with strategy for the alliance/partnership and overall relationship dynamics. Members should have decision-making authority and ultimate responsibility for the success of the alliance, as well as serve as the executive sponsor for the alliance in their respective organizations. The JOC should meet more frequently than the governance committee as members are charged with oversight and delivery of day-to-day operations, operating plans and milestones. Members of this committee should have the technical knowledge to lead the operations and decision-making authority at an operational level to minimize elevation of disputes to the governance team.

If the deal environment changes substantially, such that mutual interests become misaligned or issues in development change the scope of the original deal, both parties may need to consider renegotiation of terms.

Biotech Summary

Large pharma and biotechnology companies have profited from mutual partnering relationships in the past and large pharma is spending a significant amount of cash to access early-stage biotechnology assets. Alternative business models are now being employed to create partnering solutions whereby a biotechnology company can share risk while retaining control of its assets. In summary:

- Biotechs have different needs for partnerships, depending on their level of maturity.
- Direct equity investments can be a solution for some biotech companies seeking to gain infrastructure and intellectual capital and retain intellectual property.

- Single product risk, associated with biotechnology companies, can be mitigated by forming a portfolio.
- Biotechnology companies gain intellectual property and clinical drug development experience by choosing the right partner.

Innovative Partnering Strategies

A partnering solution for large pharma companies that is often discussed but is difficult to execute (in the author's opinion) is the creation of a Special Purpose Entity (SPE) to house a portfolio of products.

The key benefits of creating an SPE are:

- Accounting treatment may be transferred from the partner's parent company to the SPE.
- Larger sums of capital can be obtained from disparate capital sources.
- Each capital provider agrees to the upfront requirements (e.g., returns are generally scaled to the magnitude of each capital provider's investment in the SPE).
- Larger numbers of products (or perhaps a mix of products) can be developed and moved forward (to the next value inflection point, such as PoC).

The downsides to an SPE are:

- It is difficult for potential partners to agree on the number and type of assets in the portfolio (e.g., variables such as stage of development, therapeutic area, commercialization potential).
- Progress in creating the portfolio could be delayed if larger capital providers wish to renegotiate the terms of the initial partnership started by a third-party provider with insufficient capital to fund the entire transaction.
- There may be lack of agreement on the amount of upside, what to do with failed assets, how to replace failed assets, how to define (or redefine) success, etc.
- In some countries, intellectual property may need to be transferred to the entity so the sponsor can obtain desired accounting treatment (e.g., international financial reporting standards [IFRS]). Although this may allow for the desired accounting treatment, the sponsor, in essence, loses control of its asset(s).

Commercial Perspectives

When forming a portfolio, whether by choice or by merger and acquisition, there are a number of critical decisions that are influenced by commercial considerations. Common considerations can include, but are not limited to:

- Size of the market
- Number of patients with the disease
- Number of treatments currently—and predicted to be—available over the time span of the partnering arrangement, along with patent status of current products
- Level or degree of satisfaction with the currently available treatments

- Differentiation between the candidate product and another commercially available product with the same MoA
- Probable efficacy and risk profile of the potential asset
- Ease of use (oral vs. subcutaneous)
- The internal capacity of the potential partner to market a product
- Geographic reach
- Whether the approval is conditional upon a FDA-mandated Risk Evaluation and Mitigation Strategy (REMS) program
- Formulary status and reimbursement expectations for the USA, EU, Japan, and the rest of the world, along with distribution of payer mix (e.g., if disease is predominantly seen in the elderly population, the Centers for Medicare and Medicaid becomes more important in the USA than other third-party payers)
- Distribution channel (retail pharmacy, hospital, specialty distribution, physician office)

"By choice" means that a third-party capital provider is given the opportunity to partner with a number of unencumbered products. For companies that offer a large number of product candidates for portfolio selection, the initial portfolio can be pared down to a smaller number of candidates with the right risk that makes deal palatable to the investor and the owner of the compounds.

"Merger and acquisition" means that the potential partner (e.g., owner of the assets) may have inherited a number of compounds and these candidates are not necessarily selected on a strategic, high-value basis. For example, in the oncology setting, it is often advantageous to go after the largest markets (e.g., breast cancer and prostate cancer) first, but this approach may not align with the MoA of the candidate, the commercial resources of the potential partner (e.g., sales force), or the best exit strategy for the business (sale after Phase II to large pharma). Therefore, although breast cancer and prostate cancer, for example, may be potentially more lucrative markets, one of the portfolio candidates may actually compete with an existing product in the portfolio or may end up being a "me-too" in a highly competitive market and thus, actually be less attractive from an investor standpoint (or a NPV perspective).

Common commercial questions include:

- Are the candidate assets in the same therapeutic area? If so, are they synergistic, i.e., do they treat the patient at different points in the disease spectrum, or are they competitive?
  - If a third-party capital investor wishes to create a portfolio of products in multiple therapeutic classes, in the author's experience, this diversity will require additional experts and thus add cost to the due diligence exercise.
- Are the candidates in different stages of drug development?
  - Return or success criteria must be modified for each candidate.
  - If so, it may make it more complicated to define a milestone for Phase I asset success vs. Phase II (PoC) success.
  - The range of probability of success (PoS) may make the cost of capital too high for the potential partner.

- Does the return structure include a commercial element? If the third-party capital provider is paid on milestones (e.g., the entire amount based on regulatory approval alone), it may not be important, from the investor's standpoint, to conduct commercial due diligence.

Direct Equity Investments

One solution that has proven successful for earlier stage biotechnology partnering deals is a direct equity investment (DEI). These smaller investments usually require a lead investor to validate due diligence and minimize due diligence time and cost for co-investors.

The author has experience in performing due diligence in conjunction with, or on top of, others' due diligence activities (mostly for independent validation purposes) for multiple, small cap, early stage companies, where the capital provider puts a small amount down on a lead product with the hope that it will progress to the next stage (and thus increase the value of the company) or sell it to a large pharma company as a viable exit strategy. For this goodwill (and demonstration of expertise), many companies will partner with the capital provider such as a CRO, if it can provide drug development services such as the conduct of a trial, particularly if those services are offered at a preferred vendor rate.

This approach has two main drawbacks: (1) The company may not have enough money to progress to the next stage; and (2) The drug may fail. One large upside for a potential partner could be the valuation and exit reward for a biotechnology company purchased by a large pharma company. An upside to partnering with a CRO is the additional benefit of access to technical expertise and global clinical trial contract services.

Some companies have enough DEIs in a portfolio that, in the event several companies fail to progress their compounds, others will advance (and thus, increase in value) or be purchased to make the initial investments palatable and obtain a reasonable return on investment.

Summary

Investing in pharmaceutical products is a risky business. One way that single-product risk can be mitigated is by forming a portfolio of products. Individually, these products may have a high potential for technical failure, but grouping them with other compounds can increase the chance for single or multiple successes. Portfolios are generally limited with regard to the number of assets, the stage of development of each candidate, and the amount of capital required to advance each candidate to the next stage of development or sale (e.g., out-licensing as an exit strategy).

Due diligence, performed on behalf of the capital provider, should investigate the strengths and weaknesses of each candidate in the portfolio to understand the likelihood of loss of capital. If the results are unattractive, the deal should be modified or, if that is not possible, the candidate should be discarded and a substitute identified. In general, the greater the risk taken, the greater the expected return.

The optimal number of candidates can be defined using a variety of methods ranging from the approximate probability of a return on investment (e.g., the Solvay case study), based on certain milestones (e.g., the Eisai case study) to a full statistical model using a Bayesian approach.

Single sources of capital are easier to work with than syndications when funding a portfolio. Deal complexity increases with a diversified portfolio in multiple therapeutic areas at different stages of drug development (and increases the variety of due diligence experts needed to understand the technologies/candidates) and generally decreases when all products are in the same therapeutic area and stage of drug development. Less risk is usually associated with later stage compounds, as more is known about them from both safety and efficacy perspectives, and for this reason, these are probably the most attractive and most difficult to find.

If set up correctly, the partnership should not end at execution. It is critical that after the deal is executed, the portfolio partnering arrangement be optimally managed. Each company must assign a dedicated executive to manage the day-to-day operations of the product partnering arrangement, and may also form a governance committee.

If the deal environment changes substantially, such that mutual interests become misaligned or issues in development change the scope of the original deal, both parties may need to consider renegotiation of terms.

Due to the ongoing need for capital and drug development expertise, portfolios will continue to offer an attractive partnership/investment option. Companies that wish to partner on portfolios of products and have access to both capital and scientific and clinical trial (e.g., operational) expertise will be at an advantage over those lacking such resources.

Case Studies

Eisai Portfolio

In the final quarter of 2009, Tokyo-based Eisai Co. Ltd. announced it would partner with US-based NovaQuest—then part of Quintiles—on the development of six cancer drug candidates from its R & D pipeline[1, 2]. The partnership between Quintiles and Eisai is still ongoing today.

[1]*Jiho* (electronic newsletter), Quintiles to offer more risk-sharing models: Dr. Winstanly, *Pharma Japan*, 4 January 2010; Factiva Inc., 2010

[2]Drew J (2009) Quintiles teams with Eisai on cancer drugs. Triangle Bus J. http://www.bizjournals.com/triangle/stories/2009/10/26/daily62.html. Accessed 3 June 2013

The agreement called for NovaQuest to co-fund 11 PoC studies on the six cancer drugs. Quintiles' CRO business agreed to run 11 Phase II studies on the candidates and provide capacity expansion through collaboration (e.g., expertise, infrastructure, and funding). In addition to the 11 trials that Quintiles would be conducting, Eisai was to run 18 trials on the six drug compounds. The trials were designed to test the effectiveness of the drugs on a number of different cancers. The deal offered an opportunity to accelerate the development of the cancer drugs in Eisai's pipeline. The development plans were created through a robust governance structure aimed at high levels of collaboration.

The development of anticancer agents is expensive because of the need to conduct a large number of clinical trials to develop multiple indications for a single compound. R & D resource limitations make this process difficult. A risk-sharing and reward model allows pharmaceutical companies to develop multiple anticancer candidates at the same time by minimizing initial investments. Diversifying development and funding via a partnership with a CRO enables companies to shorten development and maximize sales.

The deal leveraged Quintiles' strength in cancer-drug testing. Between 2000 and 2009, when the agreement with Eisai was announced, Quintiles conducted 640 oncology studies involving more than 131,000 patients at nearly 20,000 sites in 68 countries.

According to Mr. Hideki Hayashi, senior vice president and chief product officer for Eisai, "We are maximizing the potential of Eisai's oncology compounds. I am pleased that Quintiles and Eisai share the same goals and our incentives are aligned for speed, quality, and efficiency. We will explore multiple indications in parallel so that we can deliver our compounds as fast, widely and appropriately as possible for cancer patients' benefit"[3].

This case study was created using only publicly available information as noted.

Solvay Portfolio

Solvay Pharmaceuticals had a robust pipeline of products in Phases II and III but lacked both the global infrastructure and financial resources to develop these products in a timely fashion. NovaQuest—then part of Quintiles—and Solvay formed a multiphase strategic and co-investment alliance.

Step one was a master services agreement giving Solvay access to state-of-the-art clinical trial execution by Quintiles at competitive, volume-dependent prices. The relationship simplified and accelerated study start-up, reducing the CRO's learning curve and the delays often caused by contract negotiations for each new product. This streamlined operational relationship significantly enhanced Solvay's development capabilities.

[3]Quintiles Announces Strategic Alliance to Advance Eisai's Oncology Pipeline (2009). Obesity, Fitness and Wellness Week. http://www.newsrx.com/newsletters/Obesity,-Fitness-and-Wellness-Week/2009-11-21/331121200930800W.html. Accessed 3 June 2013

Step two was a services partnership to help fund Solvay's Phase II portfolio. Under the agreement, NovaQuest provided $25 million in professional fees for up to 10 proof-of-principle studies over 3 years, making it possible for Solvay to process more assets through its Phase II pipeline. Solvay, in exchange, made milestone payments to NovaQuest for each compound that reached positive proof of concept in Phase II and moved into further development. This helped to match Solvay's out-of-pocket costs with the increased value of its assets. Each partner bears about 50 % of the normal costs and risks for outcomes.

With this customized operational and co-investment relationship in place, the partners were firmly aligned to optimize the speed, cost, and quality of clinical development programs across Solvay's portfolio. From 2001, when the deal was inked,[4, 5] until 2007 (the date of the published report upon which this case study is referenced), the alliance achieved the following results:

- The completion of three Phase III trials.
- The shift of two compounds from Phase II to Phase III testing 1 year earlier than expected, which holds the potential for millions of dollars in additional patent-protected sales if either product is approved.
- The quick production of conclusive Phase II data. (Solvay then stopped the development of the two compounds, freeing resources to work on others).
- The receipt of a $298 million contract from the US Department of Health and Human Services to develop cell-based influenza vaccines, with Quintiles to conduct nine clinical trials.
- The completion of the influenza "annual update" studies that secured yearly marketing rights for Solvay's traditional egg-derived vaccines.

Solvay Pharmaceuticals and Quintiles used the balance scorecard kit[6] to manage their alliance and together reduced the total cycle time in clinical studies by 40 %.

This case study was created using only publicly available information as noted.

Hospira Portfolio

As increasing numbers of highly successful biologics come off patent, biosimilars are a promising area for investment, offering growth potential that is lacking in many other areas of the biopharma market. Since it is projected that biosimilars will

[4]Perkins T (2007). The New Deal. Inside Outsourcing Supplement, page 36. http://www.nxtbook.com/nxtbooks/advanstar/insideoutsourcing/index.php?startid=36. Accessed 25 July 2014.
[5]van Rossum B, de Witt J (2016) Multiplying development capacity: a new model. Appl Clin Trials 15(3):50–54. http://www.appliedclinicaltrialsonline.com/appliedclinicaltrials/CRO%2FSponsor/Multiplying-Development-Capacity-A-New-Model/ArticleStandard/Article/detail/310808. Accessed 3 June 2013
[6]Kaplan RS, Norton DP, Rugelsjoen B (2010) Managing alliances with the balanced scorecard. Harv Bus Rev 88(1):114–120

retain more than two-thirds of the originator biologic's price after patent expiration, there remains an enormous opportunity for investment in biosimilar development for the biopharmaceutical industry.

Pharmaceutical companies seek to partner to access incremental funding, find complementary expertise (in manufacturing, legal, development, or commercialization) or share in the risk of biosimilar drug development.

The author led the Quintiles-centric due diligence team and was involved in the investment opportunity between Hospira, a large pharmaceutical company focusing on generic injectables, and NovaQuest, a private equity firm. The deal was publically announced April 2013 and is ongoing [7].

The risk for the partnership was diluted with a combination of three products, including Epoetin, Filgrastim and Pegylated Filgrastim. Hospira is responsible for development, regulatory approval, commercialization and distribution of the products. NovaQuest will contribute up to $150 million of development funding. Hospira will fund the remaining development costs associated with the products. In exchange for the development funding, Hospira will make milestone payments to NovaQuest upon achieving the first commercial sale for each product.

Hospira will also be required to pay NovaQuest royalties based upon commercial net sales of the products. In certain instances that result in the delay or failure of the products to be marketed (other than the failure of the products to achieve regulatory approval), Hospira will be obligated to make certain payments to NovaQuest as compensation for such unanticipated events. In these circumstances, reimbursement will be made in the form of royalties related to certain sales of Hospira's on-market products. Hospira's total payments to NovaQuest inclusive of the milestones and royalties are capped at a multiple of development funding.

This case study was created using only publicly available information as noted.

Acknowledgments Special thanks go to Drs. Michael O'Kelly, Senior Director in Quintiles Center for Statistical Drug Development, Kamali Chance, Senior Director and Head, Global Biosimilars Regulatory Strategy, Global Biosimilar Unit, Quintiles, and Tony Abruzzini, Vice President of the Center for Integrated Drug Development, Quintiles, for their editorial reviews, and the Regulatory Affairs Professionals Society, which previously published key concepts elaborated upon in this chapter.

---

[7]Hospira and NovaQuest Co-Investment I, L.P. Enter Into Collaborative Arrangement. NovaQuest Capital Management Web site. http://www.nqcapital.com/2013/04/hospira-and-novaquest-co-investment-i-l-p-enter-into-collaborative-arrangement/. Published 29 April 2013. Accessed 22 May 2013

Chapter 5
Challenges of Portfolio Management in Pharmaceutical Development

Charles Persinger

Portfolio Management in Big Pharma

Pharmaceutical companies, especially "Big Pharma," have always needed to manage their portfolio of assets in development and have leveraged traditional portfolio management approaches for over 20 years. The need for and benefit from portfolio management has increased over those 20 years as pharmaceutical development has become substantially more expensive; the scientific, regulatory, and commercial uncertainties have increased and cost pressures and resource constraints have placed a premium on productivity. The industry has found many aspects of portfolio management approaches to be useful, including:

- Prioritizing and choosing investments when an abundance of opportunities (potential investments) conflicts with resource constraints
- Identifying gaps between corporate goals and portfolio output, thus directing internal development efforts and business development
- Helping balance investments between different trade-offs (e.g., risk vs return) and different strategic categories of investments (e.g., near-term vs. long term investments, across different therapeutic areas)

This chapter focuses on the first of those aspects—prioritizing and choosing investments in a resource constrained environment. This chapter also focuses on portfolio management (how investments are made) and not the portfolio strategy (including goals and objectives), which should guide the purpose of and direction for the portfolio.

C. Persinger (✉)
Eli Lilly and Company, Indianapolis, IN 46285, USA
e-mail: cpersinger@lilly.com

Z. Antonijevic (ed.), *Optimization of Pharmaceutical R&D Programs and Portfolios: Design and Investment Strategy*, DOI 10.1007/978-3-319-09075-7_5

Portfolio management approaches for choosing between alternatives in a resources constrained environment have been used across many industries, with the financial services industry often leading the way. Within the pharmaceutical industry, these approaches were discussed in the 1998 Harvard Business Review article outlining "How SmithKline Beecham Makes Better Resource-Allocation Decisions" [1]. That article highlights many of the aspects of traditional portfolio management:

- Choices of include (fund) or exclude (not fund) for individual development projects
- Potentially choices within "fund" of different levels of funding
- Quantified value from each investment alternative
- Quantified cost (resource need) from each investment alternative
- A paradigm for ranking or prioritizing investment alternatives to deliver the highest "portfolio value"

These approaches have been widely used, often leading to periodic (annual or semiannual) portfolio reviews where the portfolio of projects to be funded are reviewed and chosen for the next period. This has required companies to need to develop approaches to consistently value projects, including forecasting revenue, assessing technical, scientific and regulatory risk, developing timelines, and estimating costs in a consistent manner so that projects can be directly compared against each other.

Best practices have been developed to generate, collect, and analyze this portfolio data. Standardized software is available and many companies have dedicated individuals or a group within their company that is responsible for this activity.

An Example

This traditional approach might look something like the following:

The first step is to gather available investments (projects) and relevant information. This could happen once or twice a year to facilitate these portfolio reviews and decision making or this information can be collected on a relatively real-time basis by having a constant approach to collecting and updating project-level information when projects change or have new data.

Figure 5.1 represents a potential collection of this data. Projects A through F are different molecules in development and D_1 represents an alternative development plan for molecule "D." For molecule D, either D or D_1 can be pursued, but not both. "Value" is typically a Net Present Value (NPV) including the forecast revenue. This may get "probabilized" with the probability that the molecule actually makes it through various phases of development and ultimately to launch. "Cost" is the development cost, potentially for that year or over some time horizon. Information on other attributes are also collected to help with questions of portfolio balance and to understand risks, including the probability of technical success, p(TS), which is

Project	Value	Cost	Other Attributes
A	$$$	$$	
B	…	…	
C	…	…	*Therapeutic area*
D	…	…	*Phase of development*
D_1	…	…	*Level of Risk*
E	…	…	…
F	…	…	

Fig. 5.1 Project level information

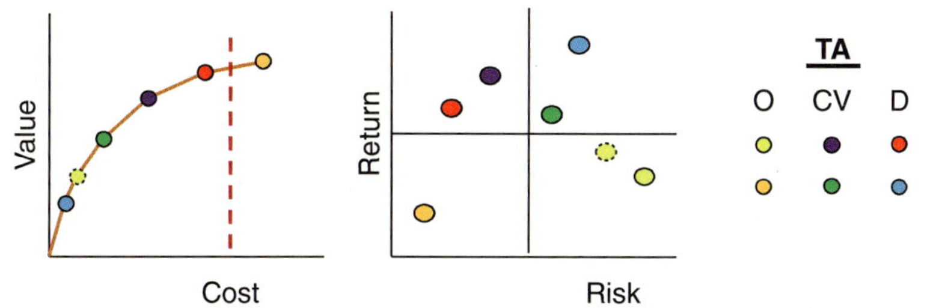

Fig. 5.2 Types of portfolio analyses/views

an assessment that the molecule will actually make it to launch (and thus deliver value to patients and revenue to the company).

After the information is collected, standard portfolio analyses are often generated to help provide insights to the decision maker(s) who are choosing between projects (molecules). These analyses may include a view of marginal cost versus marginal value, risk versus return trade-offs and balance or risk across Therapeutic Areas (TA) or other categories. Figure 5.2 shows an example of some of these types of analyses, where each dot represents one of the projects from Fig. 5.1.

Ultimately a decision is made as to which projects get funded and which do not based on the analyses, the budget and other factors that may never get quantified. This may involve creating a prioritized list of projects to receive funding (and those that do not) by analyzing the marginal costs versus marginal return, portfolio balance (risk versus return, therapeutic area, short versus long term), milestones over time (identifying any gaps), portfolio value, and other factors. This can include both quantified elements like risk and value and non-quantified elements like strategic fit. Another approach involves crafting different portfolio alternatives (i.e., different potential choices for a portfolio that include different sets of projects) and then evaluating those alternatives to see which of those portfolios looks the best on specific criteria like portfolio value. The portfolio that looks the best (or some minor tweak to the portfolio) is then selected and the projects in that portfolio receive funding and the other projects do not. Regardless of the approach, the intent is to

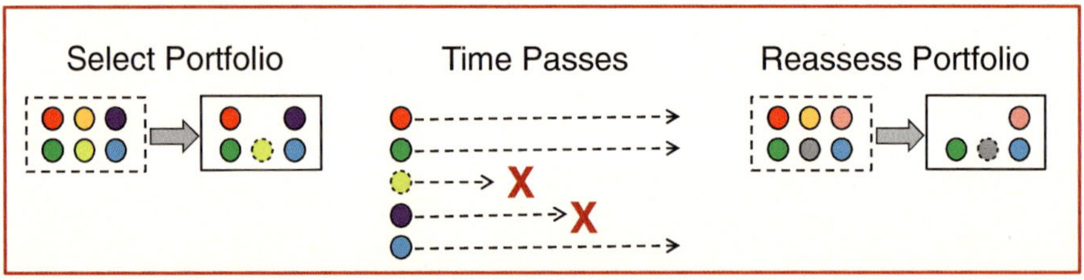

Fig. 5.3 Traditional portfolio management approach

make investment decisions (which projects receive funding and which do not) aligned with portfolio strategy, goals, and objectives and then move on to implementation and executing on those projects.

This traditional process follows a regular cadence in a company's annual calendar that defines when these decisions get made. Figure 5.3 depicts this approach where projects are selected based on analyses and approaches described above and then some time passes before the next chance to reassess and select the portfolio. At that regular interval (often annually or semiannually) the portfolio is reassessed and a new portfolio is selected. This could include stopping some projects, continuing projects, and starting new projects.

This approach depends upon:

- Project alternatives (often in vs. out, but could be different levels of investment or plans)
- Value criteria to evaluate alternatives (often NPV, cost, risk, and measures of balance)
- Credible, reliable, and consistent information for all projects

An analogy would be a personal financial investment portfolio. Typically this would start with developing a portfolio strategy, including identifying portfolio goals and objectives (e.g., saving for retirement, children's college fund, future vacation). Information is gathered about general investment opportunities types (stock, bonds, real estate, etc.). These general categories of investment would be evaluated to determine the best mix of types of investment to meet the strategic goals (e.g., if saving for retirement at the age of 25, it may involve a high percentage of stocks, but at age 60 that investment mix will likely be different). These goals, objectives, and strategic choices would define the portfolio strategy. The next step would be to create and manage the portfolio. Information would be gathered on individual investment choices (e.g., specific stocks) within the chosen general investment categories. From this, specific investment choices would be selected and the investments would be made. On a regular interval (e.g., yearly) the portfolio would be reviewed to evaluate portfolio choices in light of any changes to portfolio goals, current context (e.g., budget, age) and portfolio performance. This would result in a new investment portfolio with some investments continuing and others would being cashed out and moved to other investments. This aspect of choosing specific investments (within the portfolio strategy) and making changes over time is portfolio management and is the focus of this chapter.

The Challenge/Reality

This approach to portfolio management is useful, but faces a key challenge in the pharmaceutical industry. There are two key elements of pharmaceutical development that make this a challenge:

1. The approach requires a choice between projects (which projects get funding and which do not) at a given point in time (i.e., when the portfolio review and selection takes place). However, in a typical pharmaceutical development portfolio, projects are at various stages and points in their development and thus, there is not a chance to compare and choose between all projects (investing in the best and not funding the rest) all at the same time. Most projects, at least in clinical development, are ongoing at any point in time and cannot be stopped. For example, once a Phase 3 trial has been started (or a trial in any phase involving patients), a company is very reluctant to stop a trial for anything other than safety reasons. It is often unethical to do so and the company would not consider it, even if the portfolio management process highlighted that the company would rather divert funds from that project (i.e., stop it) to enable another project to move forward. Often, at any given point in time, there are only a few projects are at a spot where go/no-go or other investment decisions can be made on them. Certainly this is much less than half of the portfolio at any given point. Many of these commitments are 2–4 year studies. Those projects that are ongoing and cannot/will not be stopped are often called "committed" spend. Since this "committed" spend is often a large part of the budget at any given point, the degrees of freedom are greatly reduced. One approach to this might be to just hold or delay projects at a point in which multiple projects can be considered together. This sometimes occurs, especially when there are strategic considerations across those projects. However, in this industry, time is literally money (delays impact the amount of time the company can recoup its substantial expenses through revenues while under patent protection and can allow competitors to beat the product to market) and patients are ultimately impacted by any delays in development. The value of projects decline as they get delayed and thus companies are very reluctant to delay projects so that they can make decisions across all/most/many projects at once.
2. The other implication of this approach is that budgets get allocated at one (or two) times a year when projects receive funding. However, in a pharmaceutical industry, attrition is a fact of life. If a project receives funding, it may very well be terminated for technical reasons (study results etc.) sometime before the next round of decisions get made. This may free up available funds for other projects, but it makes budgeting difficult. If other projects have to wait until the next portfolio review to receive the available funding, it is another time delay and it would be preferable for those funds to be used immediately upon termination of the failed project.

The elements of this challenge are not totally unique to the industry but are exacerbated by the long development timelines, the significant attrition rate, and other major uncertainties that are inherent in the industry.

Two Different Approaches

The industry often faces these challenges with two different approaches: a portfolio-driven approach or a project-driven approach.

The portfolio-driven approach is what has been described above:

- Portfolio decisions are made that define what projects get investment
- Projects progress
- Portfolio is revisited (next budget cycle) and new decisions get made

The project-driven approach takes a different lens to this:

- Individual project decisions get made without much regard for portfolio considerations
- Projects get funding if there are current available funds and on the merits of the project
- Portfolio management is used just to report out on the aggregated view of the project level decisions that have been made (i.e., what is the portfolio view of all of the project decisions that have been made)

The portfolio-driven approach suffers from the challenges that have already been discussed. With trials underway/ongoing, it is not possible to actually choose between projects at any given time. Significant uncertainties and attrition also impact the ability to effectively execute this strategy in its purest form.

The project-driven approach has issues and drawbacks as well. If the focus is largely on project-level decisions, there is the potential to get "stuck" with an undesired portfolio. Regular portfolio reporting can help provide context for these project decisions, but without routinely taking the portfolio perspective into the project decisions, the project-level decision focus can lead to the portfolio evolving in ways that the company does not want and results in the projects driving the company strategy.

A not uncommon case is the following situation:

- Two projects in Phase 2 (Projects A and B)
- Phase 2 data is going to be available on both projects, but they read out 9 months apart
- The company can only afford to take one of the two forward into Phase 3 development
- The projects both have a 0.40 chance of reporting positive data (i.e., data that would lead to the ability to move the project to Phase 3 development)
- Project A will have data available first, but Project B is viewed as more "valuable" (although there is a lot of uncertainty about technical risks and value for both projects)
- Project A ends up having positive data and a decision has to be made

The project-driven approach could likely take the following path. Project A has the potential to move forward and is valuable, a portfolio lens is not considered (and "project B only has a 0.40 chance anyway…") and thus a decision is made to proceed forward with Project A into Phase 3. In a purely project-driven approach, the potential implications on Project B may not even be known or acknowledged. Ultimately if Project B is successful in Phase 2, the organization would prefer to move forward with B but cannot. Project B might get sold (out-licensed) and thus some value recouped.

The portfolio-driven approach would have likely had the decision on Project A wait until Project B data was available. In this case, Project A might have some planning work done to have it ready to go as quickly as possible after Project B data is available but it still would cause a delay (and loss of some value) to Project A if chosen. However, if Project B has positive data, the company is happy that it waited to choose the "better" of the two assets.

A "Hybrid" Approach

As is probably apparent at this point, companies often try to marry the best of each of the two approaches. The nature of the drug development requires that decisions get made (most often) on a project-level when data is available (and a decision needs to be made) on that particular project. However, unlike a purely project-driven approach, these decisions are informed by regular (or real-time) portfolio reviews and information, budget processes and portfolio-specific analyses. These provide the appropriate portfolio context for the project-level decisions to be made. In the Project A versus Project B case above, the portfolio and budget context would have highlighted the potential implication on Project B if Project A's Phase 3 development was approved, would have highlighted the difference in value between the projects and would have likely generated a conversation and decision on whether to wait for Project B or how a decision could be made that included Project B's information with minimal delay to Project A's timeline. To provide this portfolio and budget context, companies can have regularly scheduled portfolio reviews and analyses where projects get prioritized (in case decisions need to be made later), portfolio strategies get developed and topics like portfolio balance get discussed. This can (and should) be supplemented by real-time portfolio data and analyses that can be made available to support any particular project decision at any time. Budget cycles often also provide both short-term and long-term context and views of funding. This often is linked to financial reporting cycles, but also includes the company's business planning for near-term financials and long-range planning for more strategic financial considerations. Companies would use these approaches to provide the portfolio context, but unlike the pure "portfolio-driven" approach, funding decisions for the portfolio would not be made at this time unless the portfolio or financial review dictates a change. Those funding decisions again would be made at the time of an individual project decision. This interplay of these three considerations is shown in Fig. 5.4.

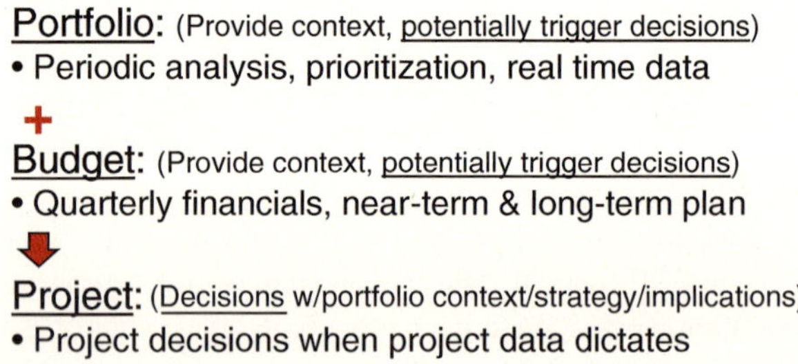

Fig. 5.4 An approach for project decisions being informed by portfolio perspectives

These strategic, portfolio considerations can be very helpful and informative when there is a clear interplay between this project's decision and another project or multiple other projects in the portfolio. These interactions can be driven by budget constraints, one project being a backup for another (e.g., multiple projects on the same target), two projects in the same disease state or many other factors. Given that these interactions are magnified across assets within a similar disease state or therapeutic area, often companies work to define portfolio strategies with a particular therapeutic area. An example would be a company pursuing multiple assets (a portfolio) in lung cancer should work to understand how they should think about the interplay, including synergies and overlaps, in projects as they make project-level decisions. In these situations, it is nice to have a clearly defined strategy of which assets get pursued and when. However, there is so much attrition and uncertainty that a rigidly defined strategy is often invalid shortly after it is crafted.

In addition to these portfolio and budget reviews and activities providing context to the project-level decisions, there are also times in which these reviews actually trigger decisions. In these cases, the portfolio and budget reviews do more than just provide context for future project decisions, they actually trigger decisions based on these strategic considerations. Portfolio reviews might identify gaps or overlaps or other strategic considerations that drive the organization to trigger specific project or portfolio decisions (selling off sections of the portfolio, in-licensing activities, etc.). Budget reviews certainly also have the potential to trigger decisions as short-term budget constraints effect immediate changes to plans or long-term views direct the company to take action.

Additional Tactics

In addition to taking this hybrid approach, additional tactics are used to help with the challenges that have been outlined. Providing more flexibility to projects through interim reviews of data and intermediate decision points can provide a better opportunity to adjust within a project to the portfolio context. For example, instead of a committing to a particular project for a 4-year study (with no chance of modifying the approach), the study could include interim looks at data for futility to enable the company to stop the study earlier. There are often implications of injecting this flexibility into a program (it may ultimately be longer, require more patients for the same statistical power, etc.), but it is often worth it for the additional program and portfolio flexibility it provides. Another approach is to offer more project-level alternatives. Teams/companies often look at only two alternatives—fund or not fund (with the "fund" alternative likely only considering an expensive "all-in" approach). It can often be helpful to develop and offer various "funding" alternatives that enable the project to move forward at lower costs or in different ways. These can include reductions in scope (maybe having a negative impact on value for that particular project) or alternate funding mechanisms (partnering, etc.). Having these options available enables the decision maker to have more (and potentially better) ways to pursue an individual project within the specific portfolio context. One final tactic is portfolio and project modeling, especially modeling the downstream implications of decisions. This enables the decision maker(s) to have a better view of the implications of a project-level decision within a portfolio context. This modeling and other analyses can help identify the portfolio implications of a particular project decision.

Summary

Tools and approaches to portfolio management have been leveraged by biopharmaceutical companies for many years to help enable development of a portfolio of projects that delivers the most value to patients and meets the company's goals and objectives. The nature of drug development however means that these approaches cannot be used as flexibly as might be desired. Since projects (investments) cannot typically be stopped mid-stream (i.e., in the middle of a clinical trial which may last years), they are not as fungible as one would like to be able to employ a standard portfolio management approach. The fact that attrition (an end to an investment) is also a frequent and sometimes randomly timed part of drug development also complicates things. This prevents a direct portfolio management approach where choices between all investments at particular time points and puts a lot of focus on individual project/investment decisions. However, it is important to recognize that, if portfolio management approaches cannot be used directly to make choices across the breadth of projects, it still has an important role to play and should not be abandoned. Companies could choose to just concentrate on making these individual project decisions in relative isolation, but that may result in decisions which harm portfolio value. Instead, a better approach is ensuring that appropriate portfolio context and implications are brought into these individual project decisions through regular portfolio updates and real-time information. Additionally, analyses that focus on the portfolio implications of particular project decisions is also helpful. These updates, information, and analyses provide the appropriate portfolio context to ensure that individual project decisions deliver both project value and advance the portfolio goals and objectives.

Reference

1. Sharpe P, Keelin T (1998) How SmithKline Beecham makes better resource-allocation decisions. Harv Bus Rev 76(2):45–57

Part III
Quantitative Methodology

Chapter 6
Impact of Phase 2b Strategies on Optimization of Drug Development Programs

Zoran Antonijevic, Jim Bolognese, Carl-Fredrik Burman, Christy Chuang-Stein, Chris Jennison, Martin Kimber, Olga Marchenko, Nitin R. Patel, and José Pinheiro

Introduction

Selecting the right dose is critical for the success of any drug development program, and for maximizing the value of a product. The high attrition rate in Phase 3 is likely due, in part, to faulty dose selection. In the new environment where payers are not reimbursing every product that is granted a marketing authorization, dose selection becomes even more important. A well-selected dose will have a better chance for a desirable risk–benefit profile and thus increase the chance for the product to be reimbursable, if it is also judged to be cost-effective. It will also result in improved

Z. Antonijevic (✉) • J. Bolognese
Cytel Inc., 675 Massachusetts Ave, Cambridge, MA 02139, USA
e-mail: zoran.antonijevic@cytel.com

C.-F. Burman
AstraZeneca R&D, SE-431 96 Molndal, Sweden

C. Chuang-Stein
Pfizer, 5857 Stoney Brook Road, Kalamazoo, MI 49009, USA

C. Jennison
Department of Mathematical Sciences, University of Bath, Bath BA2 7AY, UK

M. Kimber
Tessella, 26 The Quadrant, Abingdon Science Park, Abingdon, Oxfordshire OX14 3YS, UK

O. Marchenko
Quintiles Innovation, 5927 South Miami Boulevard, Morrisville, NC 27560, USA

N.R. Patel, Ph.D.
Cytel Inc., 675 Massachusetts Ave, Cambridge, MA 02139, USA

Center for Biomedical Innovation, Massachusetts Institute of Technology, Cambridge, MA 02139, USA

J. Pinheiro
Janssen R&D, 920 Rt 202, P.O. Box 300, Raritan, NJ 08869, USA

Z. Antonijevic (ed.), *Optimization of Pharmaceutical R&D Programs and Portfolios: Design and Investment Strategy*, DOI 10.1007/978-3-319-09075-7_6

patient care, and greater benefit to society. This makes dose selection the fundamental factor when it comes to the drug development program optimization.

The importance of dose selection has been recognized by the industry, and the Adaptive-Dose Ranging Studies (ADRS) working group was formed under the Pharmaceutical Research and Manufacturers of America (PhRMA) umbrella in 2005. This group published several papers in which they proposed novel dose selection methods, two of which [1, 2] provided very valuable information within the context of Phase 2b itself. What followed was investigation of impacts that dose selection strategies in Phase 2b have on the overall success of development programs [3] (from here on we refer to the latter as the "ADRS paper"). Since one's ultimate objective is to simultaneously optimize design aspects of both Phase 2b and Phase 3 trials, as a continuation of ADRS, an Adaptive Program (AP) work stream has been formed to address the issues around such optimization at a program level. AP is part of the Adaptive Design Scientific Working Group (ADSWG), now working under the auspices of the Drug Information Association (DIA) since 2010. The objective of AP is to further develop theoretical approaches for drug development optimization at the program level, and propose applications to selected therapeutic areas. Selected specific indications were Neuropathic Pain [4], Diabetes [5], and Pancreatic Cancer [6]. The first two followed the ADRS setup more closely, given that dose-ranging studies are more applicable to these indications. The application to Oncology expanded beyond a single drug, and compared different development programs that included two candidate compounds.

The same two measures have been used in all these studies: the PoS (Probability of Success) and the ENPV (Expected Net Present Value). The PoS, defined as the probability of a positive outcome, is often a key measure of the performance of different program design options. The ENPV has been selected as the primary measure, since it naturally accommodates optimization [7], providing an explicit trade-off between PoS, time delays and trial costs. The PoS itself generally increases as the sample size increases although the increase may be limited.

This article first presents a high-level overview of the ADRS paper and the three case studies [3–6] and then discusses key optimization parameters in the context of findings from these papers. In the interest of conciseness, this article does not present all details related to calculations and simulations. For that, and for a detailed discussion one should refer to the original papers.

Impact of Dose Selection Strategies Used in Phase 2b on the Probability of Success in Phase 3 (ADRS Paper)

Objectives and Scenarios

The purpose of this study was to assess the impact of Phase 2b design characteristics on:

1. The PoS in Phase 3, being defined as the probability of two successful, pivotal confirmatory trials, as usually required for regulatory approval, and

2. The ENPV of the product.

The impacts of the following Phase 2b characteristics were studied:

1. The statistical approach to dose selection; seven approaches as described in Bornkamp et al. [1]
2. The sample size used in Phase 2b (total of 150 and 250 patients);
3. The number of doses studied in Phase 2b (5 and 9); and
4. The number of doses selected to advance into Phase 3 (one, two, or two with expedited enrolment).

Endpoints

Efficacy

The primary efficacy endpoint for both Phase 2b and Phase 3 was a change in pain from baseline to Week 6, as measured by a Visual Analog Scale (VAS) on the scale from 0 to 10. The minimum clinically meaningful difference was considered to be an improvement over placebo of 1.3 units.

Safety

No safety limits were imposed onto Phase 2b simulations. In Phase 3 a toxicity penalty function was imposed. This function expressed the probability of a treatment-limiting toxicity being detected for a patient in a single phase III trial.

Drug Development Program

One Phase 2b study was followed by two identical Phase 3 studies. Two options were considered for Phase 3 studies: with one or with two doses of the experimental drug. All trials were placebo-controlled.

Simulations

The above objectives were studied under four different efficacy/toxicity dose–response profiles to assure the robustness of findings. Each efficacy dose–response profile had a corresponding safety penalty function. Comparisons of different scenarios were done by simulations of entire development programs. For the Phase 2 part, outputs and dose selection criteria from simulations performed for Bornkamp et al. [1] were used.

Dose Selection Criteria

For Phase 3 studies with one dose the dose selection criteria were targeting the dose closest to the minimum clinically meaningful improvement over placebo. For Phase 3 studies with two doses the adjacent, higher dose was also added. Dose(s) selected in the Phase 2b were subjected to predefined efficacy and safety profile, and the Phase 3 design and assumptions.

Definition of Success

The following definition of Phase 3 success (PoS) was used in the evaluation: a statistically significant improvement over placebo at the two sided $\alpha = 0.05$ level *and* an acceptable increase in number of patients experiencing serious adverse events compared to placebo in *both trials* and for the *same dose*. Probability of success for any given those can be calculated analytically, as described in the ADRS paper [3]. This was a simple and rather rigid criterion, but close enough to "real world" requirements to be useful for evaluation of different scenarios.

ENPV Calculations

The calculation of the ENPV required additional assumptions on: patent life, revenue stream, development costs, tax rate, and discount factor.

The ENPV was calculated as the product of the potential revenue times the PoS (used as a surrogate of regulatory approval) minus the development costs.

Findings

Based on the specific scenarios studied the following conclusions could be drawn:

- Study design has a major impact on the PoS (ranging from approximately 35 % to 65 %) and ENPV (ranging from approximately \$0.5B to \$1.2B). (Figs. 6.1 and 6.2)
- Two features that bring the most significant improvements in both PoS and ENPV were the application of an Adaptive Dose-Ranging Design, and selection of two doses to proceed into Phase 3.
- Larger sample size in Phase 2b resulted in smaller, yet consistent improvement in both PoS, and ENPV. An investigation over the broader range of Phase 2b sample sizes is necessary for more conclusive findings.
- The PoS and the period of exclusivity have more impact on the ENPV than the cost of the program. This is likely to be the case for any indication from which large revenues are expected. The investment into a particular program, however, usually comes from a limited source of money and it is important to analyze investments affecting multiple programs at the portfolio level.

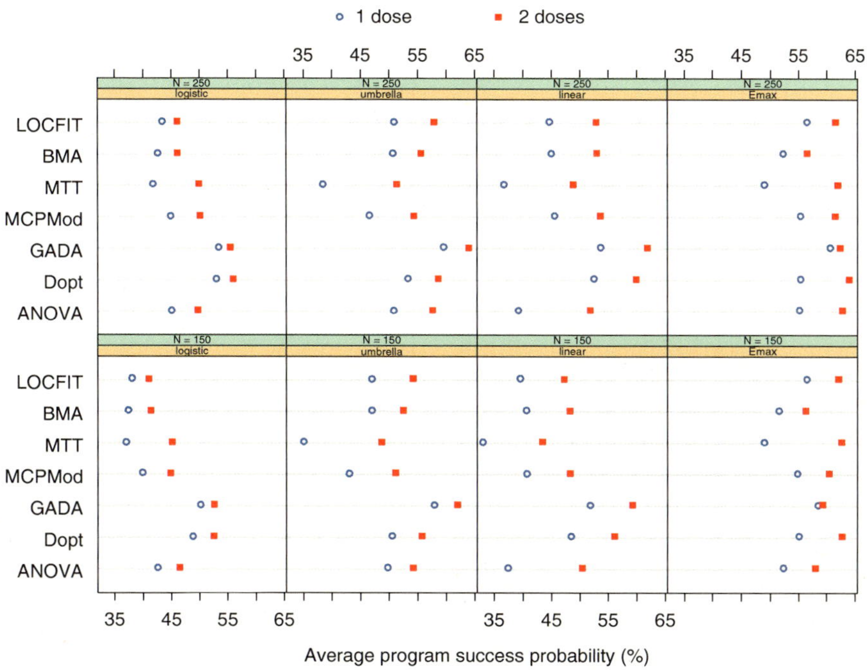

Fig. 6.1 Probability of success for the entire drug development program

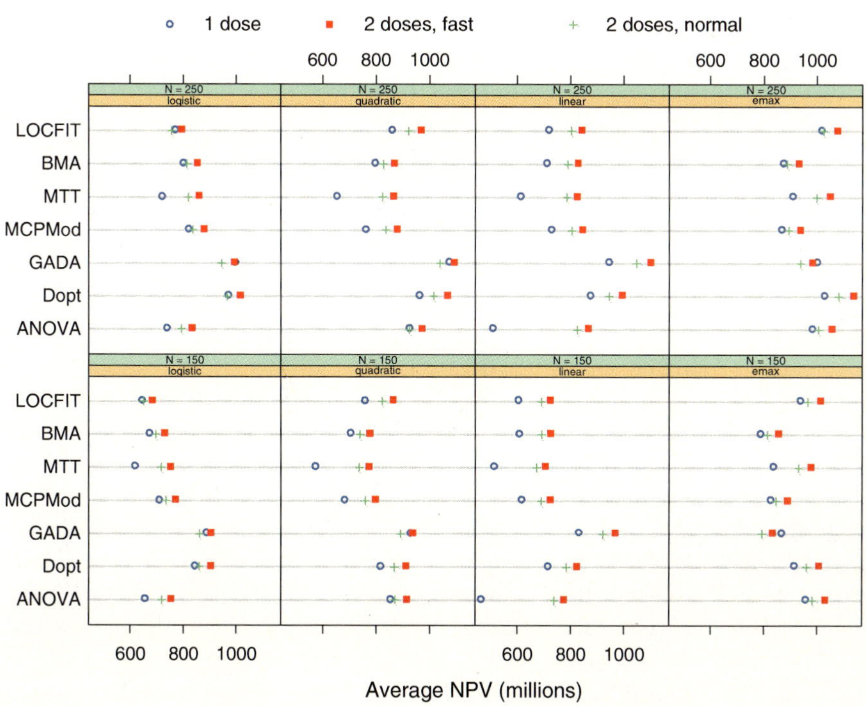

Fig. 6.2 Mean of net present values according to Phase II trial design

- The prespecified dose selection criteria did not target doses that would maximize the PoS or ENPV, as resulting selected doses were consistently lower than optimal doses.
- Simulations should play a key role in the development and evaluation of new approaches, as well as in the selection of strategies for specific development programs.

Designing Phase 2 Trials Based on Program-Level Considerations: A Case Study for Neuropathic Pain

Objectives and Scenarios

The objective of this paper was to investigate how the following factors impact PoS of the confirmatory stage and the ENPV:

1. The sample size in Phase 2b, (135, 225, 405, 540, 675, and 810 subjects total)
2. Decision rules to select one dose for Phase 3 trials, two methods were considered:
   (a) Select the dose estimated to provide efficacy closest to the target efficacy of 1 unit on the Numeric Rating Scale (NRS). The NRS has 11 categories ranging from 0 to 10,
   (b) Select the dose that will have the maximum utility expressed in 5th-year net revenue based on efficacy and tolerability.
3. The sample size for Phase 3 trials defined such that the ICH E1 requirement that 1,500 patients are treated at the dose of interest, 500 patients treated for at least 6 months, and 100 patients treated for at least 1 year is satisfied.

Endpoints

Efficacy

The proposed primary measure of efficacy, in both Phase 2 and Phase 3 trials, was change from baseline at week 12 in pain measured on the NRS. The minimum clinically important difference in mean change from baseline between the investigational product and placebo is considered to be 1 unit. However, if safety data from Phase 2b results suggest that the investigational product is likely to have a better safety profile than competing marketed products, then the target efficacy response could be lowered to 0.8 units.

Tolerability

Tolerability is measured by the probability of experiencing nuisance adverse events (AEs) commonly associated with products for neuropathic pain, which will not cause stoppage of development or prevent drug approval, but will lower the benefit–risk profile and negatively impact sales: A drug-related incidence rate of:

- Less than 0.2 is deemed better than marketed products,
- 0.2–0.3 is assumed to be similar to marketed products,
- Greater than 0.3 is assumed to be worse than marketed products.

The placebo rate for this AE will be assumed to be 0.1.

Development Program

To investigate the strategy for late stage development of a neuropathic pain medicine, it was assumed that one Phase 2 study is conducted and the results are used to determine whether Phase 3 trials should be launched. If the answer is yes, two identical Phase 3 trials would be conducted. Dose was selected based on criteria specified in Objectives and Scenarios section. All trials were placebo-controlled.

Simulations

Seven different dose response profiles were considered for efficacy. One of the seven profiles was the flat one, corresponding to a drug with 0 efficacy at all dose levels. Three levels of maximum efficacy (0.55, 1.1, and 1.65 units) and three AE dose–response profiles (low, moderate, and high) were considered. The base case was defined by a SigmoidEmax dose–response curve for efficacy with a maximum efficacy of 1.1 unit and a moderate AE profile with a maximum AE rate of 0.35 at the highest dose.

Program Success

In this case study PoS is defined as the probability that both pivotal Phase 3 trials demonstrate a statistically significant drug effect at the two sided $\alpha = 0.05$ level.

ENPV Calculations

The expected return is measured by the expected NPV discounted by a factor reflecting the declining monetary value over time. The expected return is affected by effective patent life, trial costs, relationship of efficacy and the tolerability profile of the new product (at the recommended dose), related products already on the market place, and profits of these marketed products.

Conclusions and Recommendations

- PoS is a monotonically increasing function of the Phase 2 sample size. As such the Phase 2 sample size that leads to the highest PoS for Phase 3 is 810 which is the highest Phase 2 sample sizes considered.
- For the base case, the highest ENPV under the target efficacy dose selection method occurs at the sample size of 135. For the maximum utility dose selection method, the optimal Phase 2 sample size is 270.
- Phase 2 sample sizes that lead to the highest PoS (810), versus highest ENPV (135) for the target efficacy dose selection method are on the two ends of the selected dose range. For the maximum utility method these sample sizes are much closer: all doses 405–810 for the highest PoS, versus 270 for maximum ENPV (Table 6.1).
- Since the "true" ENPV can be calculated, the "right" dose with largest "true" ENPV is known.
- Maximum utility is a superior dose selection method in all metrics considered than the target efficacy. For any given sample size, it almost doubles the PoS, more than doubles the ENPV, and approximately triples the probability of selecting the right dose and selecting the right dose or adjacent dose (Tables 6.1 and 6.2).
- The optimal sample size in Phase 2 is inversely related to maximum efficacy, and is proportional to safety severity.

Table 6.1 Comparisons between two methods to select the dose for Phase 3 development (10,000 simulations per dose-selection method and sample size)

Phase 2 Sample Size	Phase 2 Power	Prob. of going to Phase 3		Phase 3 sample size (both trials)	Total Dev. Time (Yrs)	Prob. Ph 3 Success		Expected NPV ($B)		ENPV Improvement %
		Target Dose Selection	Max Utility Dose Selection			Target Dose Selection	Max Utility Dose Selection	Target Dose Selection	Max Utility Dose Selection	
135	0.81	0.59	0.81	2880	6.7	0.55	0.81	1.22	2.03	66%
225	0.95	0.59	0.95	2800	7.0	0.56	0.95	1.16	2.30	98%
270	0.97	0.60	0.97	2760	7.1	0.57	0.97	1.15	2.32	101%
405	1.00	0.59	1.00	2640	7.5	0.57	1.00	1.10	2.25	104%
540	1.00	0.58	1.00	2520	7.9	0.57	1.00	1.04	2.13	105%
675	1.00	0.58	1.00	2400	8.3	0.57	1.00	0.98	1.99	104%
810	1.00	0.58	1.00	2280	8.6	0.58	1.00	0.92	1.85	100%

Table 6.2 Comparing dose selection based on probability of selecting the correct dose (i.e., dose 6 under the SigmoidEmax curve)

Ph2 sample size	Utility dose selection		Target dose selection	
	Probability of selecting correct dose	Probability of selecting correct or adjacent dose	Probability of selecting correct dose	Probability of selecting correct or adjacent dose
135	0.22	0.59	0.14	0.37
225	0.28	0.74	0.12	0.36
270	0.31	0.78	0.12	0.37
405	0.36	0.85	0.12	0.39
540	0.39	0.90	0.12	0.41
675	0.43	0.92	0.12	0.42
810	0.41	0.94	0.12	0.44

Optimizing Drug Development Programs: Type II Diabetes Case Study

Objectives and Scenarios

The objective of this study was to investigate the impact of selected Phase 2b and Phase 3 design parameters and decision criteria on regulatory and commercial outcomes for Diabetes drug development programs. Recommendations were made for both drug development optimization in general and Diabetes product development specifically. The impact of the following design parameters on the PoS, and the ENPV was assessed:

1. Phase 2b design (fixed or adaptive with 5 dose levels);
2. Phase 2b sample size (total of 300 and 600 patients); and
3. Phase 3 sample size (200–600 per arm).

Endpoints

Efficacy

The primary regulatory efficacy endpoint for the program was the HbA1c change from baseline which is a standard efficacy endpoint in Diabetes trials.

Safety

While not strictly a key regulatory endpoint, the incidence of hypoglycemic events are of a great medical concern, and also have a significant impact on expected revenues. Consequently they are an important component of the dose-selection criteria. In Diabetes product development CV (cardiovascular) events are safety events of the most concern. Since these events accrue at a very slow rate they cannot be included in the decision criteria regarding the dose selection or progress into Phase 3 in the same way as hypoglycemic events. As a result of this, consideration of CV events was out of scope for this paper.

Simulations

If the phase II results show that at least one dose is viable, then a single dose with the highest estimated utility is carried forward to three parallel phase III studies described below.

Table 6.3 Description of Phase III studies

Study #	Type of study	Target population	Treatment arms
1	Monotherapy study (placebo-control)	Treatment naive	Experimental drug arm(s) versus placebo versus metformin
2	Add-on combination study (active-control and placebo-control)	Add-on to metformin	Experimental drug arm(s) versus placebo versus active comparator 1 (AC1)
3	Add-on combination study (active-control)	Add-on to SU	Experimental drug arm(s) versus active comparator 2 (AC2)

SU sulfonylurea

Phase 3 Development Program

Phase 3 development program included three pivotal trials selected such that comparisons can be made against placebo and two already marketed products. It targeted three indications: monotherapy, add-on to metformin, add-on to sulfonylurea. Development program is summarized in Table 6.3.

Dose Selection Criteria

Component utility functions were created for both the primary efficacy endpoint of HbA1c and the incidence of hypoglycemia. For HbA1c, the largest utility value was 3 and was achieved when the change from baseline to endpoint relative to placebo was less than or equal to −1.3 %. The efficacy utility value gradually decreased to 0, for scenarios where placebo was better or equal to the experimental arm. The largest safety component utility value was 1 for any instances where this occurred or up to a 4 % worsening: this component utility value decreased to 0 linearly between 4 % and 30 % for greater incidence of hypoglycemia in the experimental arm. When the incidence of hypoglycemia for the experimental arm was greater than 30 % more than the placebo arm, the utility value remained at 0. The utility components for HbA1c and hypoglycemia rates were multiplied together to give the overall utility of an experimental dose.

Program Success

The following regulatory criteria has been applied:

- At least two trials have to demonstrate statistical significance for efficacy. One of these studies has to be Study #1 showing superiority versus Placebo.
- Each indication: monotherapy, add-on to metformin, add-on to sulfonylurea is approved if the above condition is met and statistically significant superiority over placebo or non-inferiority versus active control is shown in the trial for that indication.
- For HbA1c a non-inferiority margin of 0.3 was applied.

Expected Revenues

Maximum revenues of $10 billion from demonstrating superiority compared to only $1 billion from non-inferiority to active control 1, and an additional $3 billion from superiority or $1 billion from non-inferiority to active control 2. They were scaled down by penalty factors, depending on whether the drug achieves the separate efficacy goals in the separate phase III studies.

Conclusions and Recommendations

Key conclusions of this simulation study are as follows:

- Larger sample sizes in Phase 2b and Phase 3 studies provide more precise dose selection, and reduce the positive bias in the treatment effect estimate and uncertainty in estimated ENPV, within the range of sample sizes studied.
- Similar improvements are seen with implementation of an adaptive design over a fixed design in Phase 2b.
- Larger adaptive trials have identified the dose with the maximum utility dose most often, while smaller fixed designs identified this dose least often (Fig. 6.3).
- A larger number of the patients were assigned to the highest utility doses using an adaptive design compared to a fixed design for three different scenarios (Fig. 6.4).

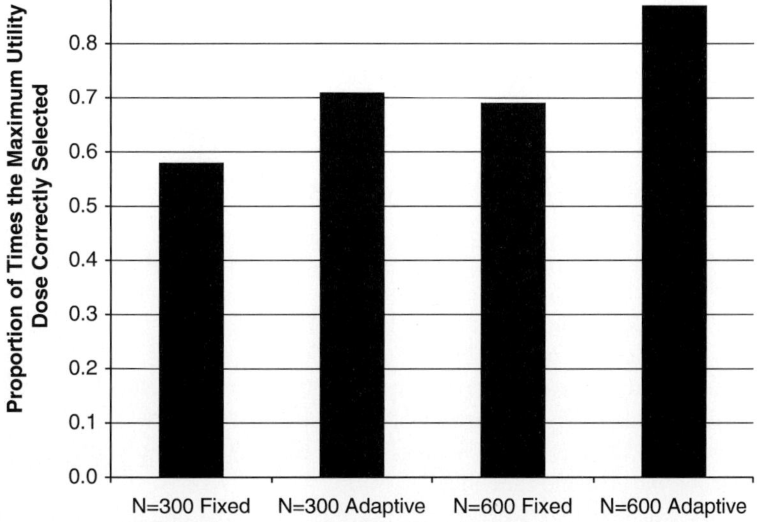

Fig. 6.3 Proportion of times the maximum utility dose is correctly identified for successful trials with the scenario where efficacy is better than expected and where the top two dose levels have high hypoglycemic episode rates

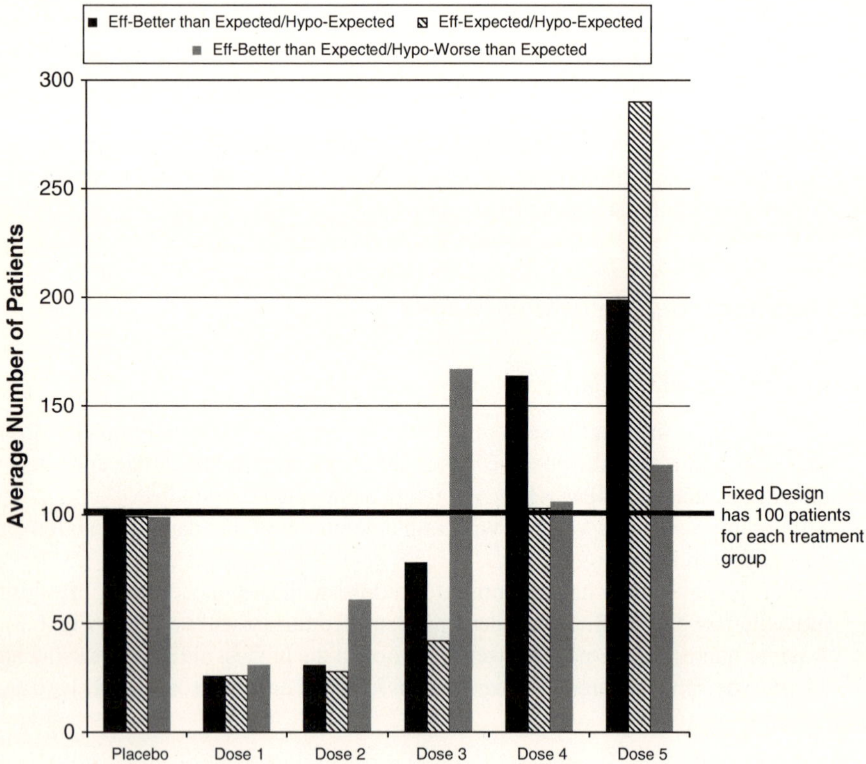

Fig. 6.4 Average number of patients: fixed versus adaptive (12 weeks with 600 patients) for three different scenarios

- Dose selection criteria have to be consistent with developers' objectives. It is a very common situation that dose selection criteria are defined by R & D teams, while one of the key objectives is to maximize the expected revenues. In order to avoid this problem we recommend closer collaboration of R & D clinical and commercial groups earlier in development [5].

Some limitations of this simulation study are:

- While some of the findings of this paper may be extrapolated to other therapeutic areas, it is recommended that simulations are conducted to support optimization of any drug development program.
- This study did not address the impact of length of development, since that would be driven primarily by a separate CV outcome study, which was out of the scope of this paper. Of note is that addressing the CV requirement would have major impact on the cost and timelines.
- Another consideration when reviewing this simulation study is that only one dose was selected to move to Phase 3 in order to limit the number of parameters that were investigated. If more than one dose is used in Phase 3, then this could have had an impact on ENPV [3].

Improving Oncology Clinical Program by Use of Innovative Designs and Comparing Them via Simulations: Pancreatic Cancer Case Study

Objectives

In this paper, authors analyzed five oncology development scenarios in the setting where a lead and a backup compound are ready for a Phase 2 trial which, if warranted, will be followed by a single Phase 3 trial. The compound here is considered in a general sense; two different drugs for the same indication or population, or two different regimens of the same drug, or two drug combinations/adds-on therapies provide an example of a potential application. Different scenarios with regard to drugs' performance were compared based on PoS and ENPV.

Scenarios

Program #1: between 2 candidate drugs (ND1 and ND2) that could be developed by a sponsor for treating pancreatic cancer one drug (say ND1) will be selected based on preclinical and phase 1 data to be examined in a Phase 2 study followed by a Phase 3 study if warranted;

Program #2: a separate Phase 2 for each of the 2 compounds, each identical to the phase-2 trial in Clinical Program #1 will be conducted and the more efficacious drug will be selected at the end of Phase 2 for the Phase 3 development if both are efficacious;

Program #3 is similar to Program #2 except that in the Phase 2 development a single 3-arm Phase 2 study to investigate both compounds and compare them with a shared control in one trial will be conducted and followed by a 2-arm Phase 3 study similar to the previous programs;

Program #4 is similar to Clinical Program #3, except that it includes a single interim analysis at which either or both ND1 and ND2 can be dropped. Simulation was used to optimize when the interim analysis was scheduled and what thresholds were used for the decision to drop an arm. Simulation and evaluation of eNPV were used to select the optimum sample size for Phase 3. Bayesian criteria were used to determine whether to drop an arm and for the "go"/"no go" decision after phase 2.

Clinical Program #5 is similar to Clinical Program #4, except that it uses more interims and a greater degree of adaptation in Phase 2 (adaptive reallocation of patients, possibility to drop an arm, early stop for futility or efficacy). Bayesian criteria were used for deciding whether and at what sample size to conduct Phase 3.

Endpoint

The primary endpoint in both Phases of development was assumed to be Overall Survival (OS), given the aggressive nature of pancreatic cancer and to simplify assumptions for the decision making.

Simulations

Hypothetical Oncology drug development programs consisting of a single Phase 2b and a single Phase 3 studies were simulated. The degree of efficacy of a new drug was expressed in terms of its hazard ratio relative to the control arm. It was assumed that the hazard ratio for each new drug ranged between 1 and 0.6 in discrete increments of 0.1. For each of the two new drugs three different settings (expressed as Bayesian priors) were specified: optimistic, uniform, and pessimistic; in terms of the probability of each discrete hazard ratio.

PoS

The success of in Phase 2 was assessed at the applicable level of significance, as described in the paper. For programs # 2, 3, 4, and 5, if both compound tests were significant, then the compound with a smaller *p*-value advanced. The success in the Phase 3 trial was defined as statistically significant comparison of selected compound with the control arm at the one-sided 0.025 level.

ENPV

The ENPV was calculated as a probability-weighted average of the associated costs and revenues, taking into account the probability distribution of the underlying degree of efficacy of the new drugs and taking into account the probabilities of reaching each stage of clinical development and product approval (conditional on the degree of efficacy). Detailed assumptions for cost and revenue parameters are presented in the paper [6].

Conclusions and Recommendations

Under the assumptions that, on average, both compounds are somewhat efficacious, four programs performed as following in terms of PoS as well as the ENPV:

- Program 1 was the worst, as expected, since it includes a single compound selected based on limited clinical data, while other programs collect Phase 2 data prior to making selection between two compounds.

Table 6.4 Overall probability of clinical program success

	Pessimistic	Uniform	Optimistic
1-Drug program (Program #1)	0.22	0.35	0.47
Parallel Ph2s (Program #2)	0.29	0.43	0.53
Combined Ph2 (Program #3)	0.34	0.50	0.61
Adaptive (Program #4)	0.34	0.50	0.62
Adaptive (Program #5)	0.40	0.57	0.69

Table 6.5 Overall eNPV

	Pessimistic	Uniform	Optimistic
1-Drug program (Program #1)	181.41	324.83	434.78
Parallel Ph2s (Program #2)	247.20	417.80	495.85
Combined Ph2 (Program #3)	322.94	529.99	625.13
Adaptive (Program #4)	313.75	514.41	605.37
Adaptive (Program #5)	340.88	552.88	651.14

The NPV is in units of millions of USD

- Therefore, if there is no sufficient evidence to demonstrate that one candidate is better than the other, one is better off to develop both as done in Program # 2.
- The most striking improvement was from Program 2 to Program #3. Even though the increase in probability of success was modest, the impact on ENPV was huge. This could be a result of time/resource savings using a shared control in a single study compared to two studies.
- Program #3 is the second best program. Program #4 did not bring much improvements compared to Program 3; in fact it performed worse in terms of the ENPV perhaps due to add-on operational costs and duration.
- Program #5 is the best choice. It has important adaptive features that separate it from other alternatives. Both the probability of success and ENPV are the highest for Program #5, justifying the use of adaptive design for Phase 2 in the setting where there are multiple compounds to develop simultaneously. Added costs associated with complex adaptive design were well offset by gain in efficiency translating into greater ENPV for this program.
- Please refer to Tables 6.4 and 6.5 for comparisons of above programs based on PoS and ENPV respectively.
- In Programs 4 and 5 the size of Phase 3 study depends on the outcome of the Phase 2. A great outcome of Phase 2 is likely an indicator that a moderate sample size is likely to produce adequate power or probability of success. On the other hand, if the Phase 2 result is less impressive, there is a high chance a large Phase 3 is needed to ensure enough power. This approach is essential to optimization at the program level, and it contributed to efficiency of Program 5.
- Only Phase 2 designs were varied in this research and Phase 3 was assumed to be a simple fixed 2-arm design trial. The approach could be extended to varying

Phase 3 designs as well; however, it introduces additional complexity and will perhaps be addressed by authors in the next paper.
- No clinical program is optimal for all conceivable scenarios and there is no one simple metric by which they can be compared. In practice, one should always come up with a few alternatives and do detailed analysis under scenarios applicable to their particular situation.

Discussion

Drug Development Scenario Comparisons

One of the most important messages resulting from these studies is that there is not one solution for all possible development scenarios. Different programs are studying different indications, sponsors have different objectives, are willing to accept different levels of risk, or are using different options to finance their trials. We recommend that every development program should specify a series of alternative program scenarios and compare them via trial simulation or analytical analysis as applicable. While the analytical solutions should be a preferable approach whenever possible, the reality is that optimization at the program level is very complex, and may require simulations of entire programs as illustrated in these studies.

Another consistent observation from these studies is a high correlation of conclusions based on PoS versus ENPV. Almost always programs that performed best in terms of PoS also performed best in terms of the ENPV. One exception is the Neuropathic Pain study [4] when target efficacy is used as the dose selection criteria. It has, however, been demonstrated in that study that this method is neither a reliable one, nor a desirable one when it comes to optimizing drug development programs. This high correlation between PoS and ENPV is partly due to all indications studied having relatively large expected revenues. In case of indications with smaller expected revenues the correlation may not be that high. Regardless, these results should send a strong message to developers that their primary focus should be on the success of drug development programs, rather than the race who will get to submission first. After all, their competition may be failing as well, particularly if they also are giving the speed of development priority over the sound drug development, as there is a trade-off between speed of development and accuracy.

It has to be noted that these simulation studies were conducted at the time when reimbursement was less of an issue. Therefore the regulatory approval was considered to be the last hurdle before one can start accruing revenues. While that is still technically the case, in the current environment revenues will experience a major hit if the drug is not reimbursed. We recommend that reimbursement is included more specifically as a parameter in future simulations of drug development programs. This brings back the issue of speed of development. In the future, a better differentiated product will be more commercially successful, regardless of who gets to the submission point first. There are a number of ways to assure differentiation, one obviously being a careful dose selection. Another one is selection of a biomarker subpopulation that is more likely to differentiate with its benefit–risk profile than the whole population. Discussion of such designs is beyond the scope of this chapter, but is included elsewhere in this book.

Dose Selection Criteria

Selection of the optimal dose has two main components. The first component is the criteria for selecting target dose(s), and the second component is the method that best identifies target dose(s) in terms of maximum precision and minimum bias. In this section we discuss the former. The accuracy of dose selection is regulated by the study design and Phase 2b sample size and is discussed in the following sections.

Two methodological papers mentioned [1, 2], focused only on the target dose, and did not address the optimization. The ADRS paper [3] used the dose selection criteria that was not in concordance with optimization criteria, and as a result of that selected doses were consistently lower than the optimal dose.

The neuropathic pain paper [4] made very nice and thorough comparisons of two dose selection methods, one based on the target efficacy and the other on the maximum utility. The maximum utility far outperformed the target efficacy as the dose selection method in all parameters measured. This finding is a very clear demonstration that focusing on sequential, trial by trial based targets may not result in the most efficient drug development even if the best methods are used for individual trials. Rather, decision criteria should be defined in such a way that outcomes related to desired output of the entire program are maximized.

The diabetes paper [5] did something slightly different. It attempted to mimic a typical development process in which the utility function is specified by a clinical team and expected revenues are projected by a commercial development team. While two utility functions (clinical and ENPV-based) looked similar, there were differences in which development options maximized clinical utility, versus which development options maximized the ENPV. Since the clinical utility function was applied to select the dose, the implication was that the best dose selection method did not always maximize the ENPV. This example emphasizes one of the main messages of this book as a whole, that there has to be an integrated rather than siloed approach to drug development.

Oncology development program is different from programs in other therapeutic areas. While dose-ranging studies are conducted as Phase 2 studies in most therapeutic areas, dose-ranging or rather dose-escalation studies are Phase I studies in Oncology with an objective to select the maximum tolerated dose or the optimal biologic dose. In the Oncology paper [6] instead of dose selection, the drug selection was performed and the decision criteria were very simple, not much of the factor. In Program #1 there was no decision besides at the end of Phase 2 whether to proceed to Phase 3 development. In other four programs decision was based on comparison of two compounds, and one that performed better in terms of efficacy even by the slightest margin would be selected. The main difference between five programs was the amount of information available at the decision point. We can conclude that a better informed criterion results in the better program success.

Study Design

Adaptive Design

With emergence of adaptive design, the selection of statistical methodology becomes one of the most critical components of study design. The flexibility of adaptive designs makes them particularly suitable for the exploratory stage of development. Adaptive design allows for key design parameters to be changed during the trial based on the data observed during that trial, and as such is a natural fit for the exploratory stage of development. For example, in Phase 2b trials ineffective doses can be discontinued, new doses added or the randomization allocation ratio adjusted in favor of doses that demonstrate better safety/efficacy during the trial. As a result, over the course of a trial a larger and larger proportion of patients would be randomized to doses with better safety and efficacy characteristics. Findings of the previously mentioned papers was that adaptive design outperforms other Phase 2b designs, whether by selecting the right dose [1, 2, 5], right development program [6], maximizing the Phase 3 PoS [3, 6], or ENPV [3, 5, 6].

The utility of adaptive design does not stop in Phase 2b, but has its application in Phase 3 as well. Phase 3 adaptive designs were not studied in these papers, and are therefore beyond the scope of this chapter, but we describe several key features/advantages of these designs. They can be very useful in situations where after the exploratory stage development there is still:

- A residual uncertainty regarding the treatment effect. In this case there are two options.
  - One is the traditional group sequential design (GSD). With this design the maximum sample size usually targets the minimum meaningful clinical effect, but there are options for closing the study early at an interim analysis for futility or to claim efficacy.
  - The other is the unblinded sample size reassessment (SSR). This design initially targets a more optimistic treatment effect, but has an option to increase the sample size based on the observed treatment effect at an interim analysis, and as such "salvage" a trial that would have otherwise been borderline negative.
- A residual uncertainty regarding the optimal dose. In this case the Phase 3 trial can start with several doses, and use an interim to select dose(s) for the second stage of the trial where the efficacy of selected dose(s) would be confirmed.

- No sufficient evidence that a Biomarker-based subpopulation risk–benefit profile is better than that of the full population. In this case one can select an adaptive enrichment design which identifies predictive markers and confirms efficacy within a single trial.

Phase 2 Sample Size

Two of the papers [3, 5] considered only two options for the sample size, and within the range studied the larger sample size performed better judged on outcomes of interest. The question of Phase 2b sample size was thoroughly addressed only in the neuropathic pain paper [4]. Findings of this paper are quite interesting. The PoS as a function of the Phase 2b sample size was monotonically increasing, while that is not the case for the ENPV where there is an optimal Phase 2b sample size (270) that maximizes it. These findings confirm previous speculations [7].

Let us use this information to further discuss sample size calculation for Phase 2b trials. It is a much more complex concept than calculating sample size required for a confirmatory trial, where it is based on the power to reject the study hypothesis assuming a treatment effect of interest. As previously explained, the Phase 2b sample size will impact the precision of dose selection, with larger sample sizes improving precision while increasing the cost and length of drug development. In the context of program optimization the Phase 2b sample size should be selected such that it maximizes the outcome of interest. The recommended procedure to calculate the sample size would be as described in the neuropathic pain paper: consider a wide range for Phase 2b sample size options, run simulations, and find what sample size maximizes the measure of interest.

Calculating sample size based on pair-wise comparisons that has been traditionally applied in drug development would not be recommended based on results of these studies, since design based on pair-wise comparisons did not perform as well as model-based adaptive designs [3].

Other Design Parameters

Number of doses to be included in the Phase 2b was studied in two papers [3, 4]. Findings of the ADRS [3] paper were that designs with smaller number of doses (five) performed better for non-adaptive designs, while for adaptive designs the number of doses in Phase 2b made very little difference. The findings of the Neuropathic Pain paper were similar, with smaller number of doses (four) performing slightly better than the larger number of doses (nine) for non-adaptive designs. These findings suggest that a larger number of doses should be considered selectively, and only if an adaptive Phase 2 trial is planned. The "optimal" number of doses would best be determined by running simulations.

Only one paper [3] studied the number of doses to proceed into Phase 3, but improvements of Phase 3 with 2 active doses versus 1 active dose were striking.

This finding is also quite intuitive, since addition of second dose would definitely de-risk the trial avoiding situation where a single selected dose could be either too low in terms of efficacy, or too high when it comes to safety.

Looking Beyond a Single Compound

The oncology paper [6] is taking Phase 2 considerations to another level as it addresses the compound selection, while other papers are assuming that the compound/indication of interest has already been selected. The decision criteria are also somewhat different from other papers, among other reasons because the decision is to be made to select one out of only two options. Results clearly illustrate various possibilities for improvement over the basic design. The more options for adaptation considered, the better the outcome. The best performing program design included adaptive features such as adaptive randomization, early stopping for futility and efficacy, and Phase 3 sample size which depends on the outcome of the Phase 2.

In drug development often a number of issues need to be addressed simultaneously, such as indication selection combined with biomarker and/or dose selection. Heterogeneity of disease is a widely acknowledged problem in cancer, and very difficult to address in the course of a clinical trial. It is the "next big problem" that oncology trial design has to tackle; perhaps designs used in BATTLE, and I-SPY 2 trials (www.ispy2.org) that pair oncology therapies and biomarkers point the way. This setup could be expanded into addressing several development questions simultaneously. Many companies have multiple assets at the same stage, or different stages of development. These situations inevitably lead into optimization at the portfolio level, which is the main topic of this book.

Conclusions

While four papers looked at the drug development from different angles, and considered different indications there are some common findings, based on which we can conclude the following:

1. Selected design and decision criteria have a great impact on PoS and ENPV.
2. It is therefore recommended that multiple development options are quantified and compared based on the outcome of the interest. Given the complexity of drug development, simulations are the best way to enable comparisons.
3. Dose selection has two key components.
   (a) The criteria for selecting target dose(s), and
4. Some dose-selection criteria, such as target efficacy value, are not optimal by their nature. Doses with efficacy better than target efficacy and acceptable safety perform much better than the target efficacy dose.
5. Optimal drug development can be accomplished only when dose selection criteria are consistent with ultimate program objectives.

   (a) The method that best identifies target dose(s) in terms of maximum precision and minimum bias.
6. Adaptive designs in Phase 2b far outperformed fixed designs.
7. Larger Phase 2b trials always outperform smaller trials in terms of the PoS. If the ENPV is the selected outcome, then there would be a corresponding, optimal Phase 2b sample size due to the trade-off between increased cost and longer development.
8. The ENPV is a very well-suited measure for drug development optimization at the Phase2b/Phase 3 stage:
   (a) With increased investments it balances between gains in PoS and increases in costs and development times. As such it provides an optimal solution for parameters such as the sample size.
   (b) At this stage of development the ENPV is a relatively stable measurement, and it can be updated based on learning that are internal or external to the given product development.
   (c) It is a well-established parameter for internal decision making.
9. Considerations need to be given to the optimization at the portfolio level, since decisions at individual program levels are interrelated due to budgets usually being imposed at a higher, portfolio level.

References

1. Bornkamp B, Bretz F, Dmitrienko A, Enas G, Gaydos B, Hsu CH, Koenig F, Krams M, Liu Q, Neuenschwander B, Parke T, Pinheiro J, Roy A, Sax R, Shen F (2007) Innovative approaches for designing and analyzing adaptive dose-ranging trials (with discussion). J Biopharm Stat 17:965–995
2. Dragalin V, Bornkamp B, Bretz F, Miller F, Padmanabhan SK, Perevozskaya I, Pinheiro J, Smith JR (2010) A simulation study to compare new adaptive dose-ranging designs. Stat Biopharm Res 2:487–512
3. Antonijevic Z, Pinheiro J, Fardipour P, Lewis R (2010) Impact of dose selection strategies used in phase II on the probability of success in phase III. Stat Biopharm Res 2:469–486
4. Nitin P, Bolognese J, Chuang-Stein C, Hewitt D, Gammaitoni A, Pinheiro J (2012) Designing phase 2 trials based on program-level considerations: a case study for neuropathic pain. Drug Inf J 46(4):439–454
5. Antonijevic Z, Kimber M, Manner D, Burman C-F, Pinheiro J, Bergenheim K (2013) Optimizing drug development programs: type 2 diabetes case study. Ther Innov Regul Sci 47(3):363–374
6. Marchenko O, Miller J, Parke T, Perevozskaya I, Qian J, Wang Y (2013) Improving oncology clinical program by use of innovative designs and comparing them via simulations. Ther Innov Regul Sci 47(5):602–612
7. Antonijevic Z (2009) Impact of dose selection strategies on the success of drug development programs. TouchBriefings 4:104–106

Chapter 7
Using Decision Analysis to Support the Design of Clinical Trials Within a Program

Richard Nixon and Blair Ireland

Introduction

Drug development involves complex, high-value decisions with lasting consequences. These decisions are made in the context of uncertainty, with information of many different types and from many different sources. A project team making a decision regarding the development of a drug needs to structure and synthesize this information in this context and come to a consensus on the best course of action.

Team decision-making often suffers from two systemic problems. The first relates to process, where a potential cycle of decision-making in drug development is for the team to consider one preferred course of action, collect arguments and information to support this choice, and take the arguments to a senior decision-making board. This board then rejects the plan without providing an alternative, and the team has to start the whole process again, resulting in team frustration, development delays, and increased costs.

Richard Nixon and Blair Ireland are employees of Novartis. The views expressed in this chapter are the views of the authors and do not necessarily reflect the views of Novartis.

R. Nixon (✉)
Novartis Pharma AG, Postfach, 4002 Basel, Switzerland
e-mail: richard.nixon@novartis.com

B. Ireland
One Health Plaza, East Hanover, NJ 07936-1080, USA
e-mail: blair.ireland@novartis.com

Z. Antonijevic (ed.), *Optimization of Pharmaceutical R&D Programs and Portfolios: Design and Investment Strategy*, DOI 10.1007/978-3-319-09075-7_7

A second problem in team decision-making is one of cognitive bias. Decision-makers develop sophisticated heuristic capabilities that enable them to make quick decisions in the moment. Unfortunately, these heuristics often lead to poor decision-making when faced with highly complex situations. For example, satisficing involves a bias that leads us to choose a decision once it is considered good enough, rather than optimal, and the decision is reinforced by feeling good about having made the decision. Protection-of-mindset biases also can lead decision-makers to solve the problem they can rather than the problem they should.

Decision Analysis (DA) comprises a set of tools for structuring and quantitatively analyzing complex decisions. The framing and structuring elements of DA encourage decision-makers to consider a wide range of decision alternatives, and they help to overcome cognitive biases. The quantitative aspects break down complex problems into simpler problems, which can be thought through and then recomposed mathematically.

In this chapter, DA methods are applied to a case study in drug development. A drug for a chronic disease is in phase IIb of development. One of the regulatory requirements from the Pharmaceuticals and Medical Devices Agency (PMDA) in Japan is to demonstrate that the dose response in Japan is the same as that in the general population. The phase IIb study has been designed with this requirement in mind; however, recruitment in Japan has been slower than anticipated. As a result, it is predicted that the number of Japanese patients recruited when the required total sample size is attained will be insufficient to meet the PMDA's requirement. The project team needs to assess the situation and decide how to proceed with the registration strategy for Japan and the rest of the world.

Overview of the Decision Analysis Process

The six-step DA framework given by Matheson and Matheson [2] is used to address the case study. They describe the framework as a chain, because the quality of the whole decision is only as good as its weakest link.

1. *Focus on the right questions:* A *decision hierarchy* is used to identify issues to be decided now, issues that have already been decided, and issues that can be deferred until later. This helps give the decision-making team clarity on what they need to decide.
2. *Create varied, doable alternatives:* The set of alternatives usually comprises a set of interdependent decisions. A *strategy table* is used to map out possible options for each decision and link these to viable alternative strategies.
3. *Define the criteria for evaluating alternatives:* The team must decide how they value a decision. Essentially, what it is they want more of. In the example, the development program should be completed cheaply, quickly and with high probability of success. As is commonly the case, these values are in tension; for example, a strategy with a high probability of success likely will be expensive and take time.

4. *Assess information and uncertainty:* An *influence diagram* is used to assess what the team needs to know to link the decision alternatives with the criteria for evaluating alternatives. The information needed for each criterion is identified until a path is traced back to the decision or decisions. This path shows what information is needed, and the information then is gathered and used to populate a decision model. Elicitation techniques and conditional probabilities are used to gather information for which there is no direct data.
5. *Evaluate alternatives using sound models:* When the conceptual model is complete, a computer model is built. In the example, the computer model links clinical trial simulation with market models. The model allows the value of each alternative to be calculated and compared. In addition, sensitivity analysis of the underlying assumptions and uncertainty in the information can be performed. This can assess how robust a decision is to changes in assumptions and show what the main factors influencing the decision are.
6. *Stakeholders commit to implementation:* The outcome is a succinct set of recommendations that are based on a set of transparent assumptions and the results of quantitative analysis, backed up by a decision audit trail, and provide the foundation of a consensus for action by the team.

Case Study

The main purpose of a phase IIb study is to determine the dose–response relationship of a drug, and so to inform the decision about which dose and regimen is most likely to have the greatest clinical value. This "learning" stage of drug development selects a dose to be taken into phase III, which is a "confirming" stage [6].

The PMDA requires that dose response be evaluated separately in Japanese patients [4]. The guidelines are focused on phase III studies, but for this phase IIb dose–response study, the biological effect of the minimum effective dose (MED) in Japan is required to be the same as that of all patients in order for one active dose to be taken into phase III. The MEDs in Japan and in all patients are accepted as being the same if the point estimate of the ratio of the treatment effect in Japanese patients and all patients is greater than 0.5.

If at the end of phase IIb, it has not been demonstrated that the MED in Japanese patients is the same as in all patients, an option would be to take two active doses into phase III and use this to assess whether the dose response is the same in Japan as in the whole population. However, such a study would cost more and take longer compared to a phase III study with just one active arm. Furthermore, this violates the principle that the phase III study should be a confirming study, and the dose–response relationship should be learned in phase IIb.

In this example, multiple doses of a drug along with a placebo arm are included in the phase IIb study. This study, with 900 patients, has 90 % power to detect a dose-response signal in the randomized population. The study is also designed to estimate the MED required to obtain a minimum clinically meaningful difference between the drug and placebo. Of the 900 patients expected to be enrolled in the study, at least 120 are planned to be from Japan. This would allow sufficient power to demonstrate that the MED in Japan is the same as in the overall population.

The complication is that recruitment in Japan is slower than expected. This means that when 900 patients have been recruited, a smaller-than-expected proportion of them will be from Japan. At the observed recruitment rate in Japan, only 50 Japanese patients are expected to be recruited, which reduces the probability of demonstrating that the MED in Japan is the same as the MED in all patients to 20 %. Recruitment could be continued in Japan, until 120 patients are randomized, but this would delay the global submission. The core dilemma facing the team is whether to continue recruitment or to protect the global submission and perform a stand-alone Japanese program.

Focus on the Right Questions

The first stage of a DA is to understand clearly what the decision-makers want to decide. Typically, an early "core framing" meeting is held with a small group of key decision-makers, for example the medical director and project leader. Brainstorming of the key issues is performed, and these issues are sorted into decisions and information. The distinction between these is that a decision is something you have control over, whereas information is something that you could influence by a decision but have no control over. For example, the treatment effect of a drug could be influenced by the inclusion criteria of patients in the study, but it is fundamentally a state of nature that you cannot control.

Burman and Wiklund [1] make the important point that quantitative modeling should be to support decision-making; hence, without a clearly defined decision, one should not start such modeling.

A helpful tool to use for this step is a *decision hierarchy*. This categorizes decisions into "givens," "decisions for now" and "decisions for later." Givens can be decisions that are already made or restrictions due to policy or the environment. For example, drug development is performed in a highly regulated environment, and aspects of the program design that are required by the regulators fall into this category. Decisions for now are the focus of the DA. These include near or long-term strategic decisions, or near-term decisions that require significant resource commitments. Decisions for later include (1) those that involve later significant resource commitments, (2) decisions for specialists, and (3) operational or tactical decisions.

In this example the decisions that are givens are shown in Table 7.1.

The following are decisions for now:

1. Should the recruitment be extended in Japan until 120 patients are randomized?
2. If a stand-alone program is needed in Japan, what designs should be used for the dose-finding study and the pivotal registration study?

Table 7.1 Decisions that have already been made

Aspect	Decisions made
Doses	Any adaptation of the phase IIb study will keep all doses in the study. This is due to the environmental constraint that the repackaging needed to add or remove a dose would take several months and would require a substantial protocol amendment.
	If the phase IIb study does not give sufficient information for the PDMA to decide whether the biological effect of that of MED in Japan is the same as that of all patients, then a separate dose finding study will be performed in Japan.
	The global pivotal phase III study will have the same dose in all regions. If the dose needed in Japan is not the same as in the rest of the world, then a stand-alone pivotal study for Japan will be conducted.
Global phase III study	If a Japan-only extension of recruitment is performed (to enable a sufficient number of Japanese patients to be recruited), then Japanese patients can join the phase III pivotal study if the MED in Japanese patients is the same as the global population.
	The phase III program will not be put on hold for Japan dose selection.

Create Varied, Doable Alternatives

Strategy Table

There are three decisions to make. (1) Will the Japan recruitment be extended in the phase IIb study to recruit the desired 120 patients in Japan? (2) If stand-alone studies in Japan are required, how will (a) the dose-finding study and (b) the phase III pivotal Japanese submission study be designed? The next stage of the process is to define a range of options for each decision. There are two criteria for good options. Firstly, they must be doable. Whilst creativity and out-of-the-box thinking is encouraged at this stage, these must be tempered by pragmatism. Creativity can be encouraged by an outside facilitator challenging established team dogma. This is balanced by pragmatism by performing this exercise with people who understand the practical limitations of what can be done. Secondly, the options should span the full range of alternatives. They should not be variations on the same theme; rather, they should cover the range of feasible possibilities among which the decision-maker could consider choosing. The set of options does not need to be exhaustive, and it is often suitable for the number of options for each decision to be between two and five. Options do not need to be completely defined. At this stage it is enough to know only their salient features.

A *strategy table* is used to structure the decision options. For this example, Table 7.2 shows a strategy table in and details of the possible study designs are given in Table 7.3

The strategy table maps out the possible options for each decision. For this example they are as follows:

1. Phase II: Should the recruitment be extended in Japan until 120 patients are randomized?
   - Recruit 50 JP patients: Recruit until the study has 900 patients, at which point the number of patients from Japan is expected to be 50.
   - Recruit until there are 120 JP patients: Once the study has 900 patients in total, including 50 from Japan, continue recruiting a further in 70 patients in Japan until 120 Japanese patients have been randomized.

Table 7.2 Strategy table

	If need stand-alone Japan studies	
(1) Phase II	(2a) Dose finding	(2b) Phase III
Recruit 50 JP patients	None	None
		JP stand-alone
Recruit to 120 JP patients in extension	Phase IIb two active arms Phase IIb three active arms Phase IIb/III two active arms	Join global phase III at risk

This maps out the possible options for each decision. The rows in a strategy table have no meaning. Strategies are created by choosing one decision option from each column

Table 7.3 Details of the types of possible studies in the development program

Study	Description
Global phase IIb	Recruit until the study has 900 patients, at which point the number of patients from Japan is expected to be 50.
Global phase IIb with extended recruitment period in Japan	Once the study has 900 patients in total, including 50 from Japan, continue recruiting in Japan until 120 Japanese patients have been randomized
Japanese phase IIb with two active arms	A stand-alone Japanese dose-finding study with two active arms.
Japanese phase IIb with three active arms	A stand-alone Japanese dose-finding study with three active arms.
Japanese phase IIb/III with two active arms	An adaptive phase IIb/III study, which starts with two active arms, and later drops one arm.
Global phase III	Pivotal global phase III study with one active arm, including patients from Japan.
Global phase III without Japan	Pivotal global phase III study with one active arm, but not recruiting patients from Japan.
Japanese phase III	Pivotal phase III study with one active arm, recruiting only in Japan.

If the PMDA does not accept that the MED in Japan is the same as the MED in all patients, based on the phase IIb study data, then stand-alone studies in Japan will be needed. There are two decisions to make in this case.

2a. Dose finding: What should be the design of the stand-alone Japan dose finding study?
   - None: No study is performed.
   - Phase IIb with two active arms.
   - Phase IIb with three active arms.
   - Adaptive phase IIb/III with two active arms.

2b. Phase III: What study should be done for registration in Japan?
   - None: No study is performed.
   - Japanese stand-alone phase III study.
   - Join global phase III at risk, while any stand-alone Japan dose finding study is still ongoing.

Table 7.4 Table of alternative strategies

	If need stand-alone Japan studies			
Phase II	Dose finding	Phase III	Pros	Cons
Recruit 125 JP patients in core study	None	Join global phase III	"Straw man." In reality this option cannot happen as only 50 patients will be recruited in the core study. We also assume that JP is included in global phase III for an ideal base case.	
Recruit 50 JP patients	Phase IIb three active arms	Join global phase III at risk	Saves time and money with JP in global phase III.	Likely needs stand-alone study. Risk JP dose not same as RoW.
	Phase IIb/III two active arms	None	Saves time and money.	Complex study design.
	Phase IIb two active arms	JP stand-alone	Does not affect global submission.	Needs many JP patients.
Recruit to 120 JP patients in extension	Phase IIb three active arms	Join global phase III at risk	Less likely to need stand-alone JP program.	Could delay the global submission.
	Phase IIb/III two active arms	None	Less likely to need stand-alone JP program. Cheap and quick back-up plan.	Complex study design.
	Phase IIb two active arms	JP stand-alone	Less likely to need stand-alone JP program. Separate back up plan.	Needs many JP patients.

The rows in a strategy table have no meaning. To create varied, doable alternatives, strategies are created by choosing one decision option from each column. In this case six different strategies are selected for consideration. In addition, as a "straw man" the ideal case is also considered, i.e., where the initial plan to recruit 120 Japanese patients is successful. The motivation for the straw man case is to get a set of baseline costs and times to compare with the other strategies and to make differential comparisons of the other strategies to this ideal case. The ideal case, and six different alternative strategies, along with some high level pros and cons for each strategy, are shown in Table 7.4.

Decision Tree

A *decision tree* is a visual display of the different decision alternatives and possible consequences. Each node in the tree can be one of three types: A *decision node*, represented by a square; an *uncertainty node*, represented by a circle, describing a possible consequence, which may or may not be influenced by a decision; and an *end node*, represented by a triangle, at the point where the final consequences from the set

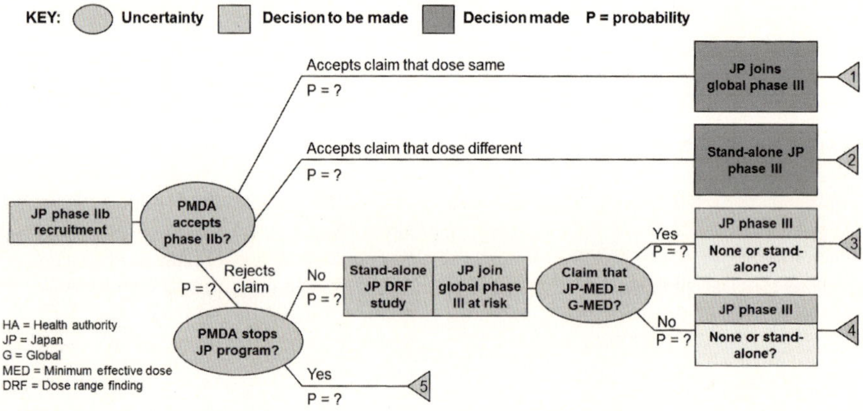

Fig. 7.1 Decision tree

of decisions are assessed. These three elements of the decision taken together are the *decision basis*; they describe what can be done, what is known, and what is wanted.

The decision tree for the case study is shown in Fig. 7.1. Not every decision option is displayed explicitly as this would result in an unmanageably large decision tree. Instead, the decision tree is collapsed but gives all the consequences that are possible following each decision. The actual decisions made will influence the probabilities of each possible consequence occurring.

The first decision is whether to extend recruitment. Following this there are three possible outcomes.

- Firstly, the study team claims that the MED in Japan is the same as the MED in all patients and the PMDA accepts this claim. If this occurs then Japanese patients will join the global phase III study. Japanese patients will only join this study when the PMDA accepts the claim.
- Secondly, the study team claims that the MED in Japan is different from the MED in all patients and the PMDA accepts this claim. If this occurs then Japanese patients will be recruited into a stand-alone phase III study in Japan.
- Thirdly, the PMDA rejects the claim made by the study team. If this occurs, it is possible that the PMDA may stop the program in Japan and no approval from Japan would be possible. However, if the program continues in Japan, a decision has to be made regarding the design of the stand-alone Japanese dose-finding study and whether Japanese patients should enter the global phase III study at risk.

In the case of the third outcome, the stand-alone Japanese dose-finding study will be powered so that the claim made from it should be accepted. There are two possible claims:

- The MED in Japan is the same as the MED in all patients.
- The MED in Japan is different from the MED in all patients.

In either of these cases, the decision about the design of the stand-alone Japan phase III submission study needs to be made.

For each strategy, there are five possible paths through the decision tree that can occur.

It could be that the *framing* steps of focusing on the right questions and creating varied doable alternatives are sufficient to bring clarity to the project team and the DA can stop at this point. Albert Einstein captures this principle elegantly when he quips, "The mere formulation of a problem is often far more essential than its solution, which may be merely a matter of mathematical or experimental skill." Once the problem has been framed, its quantification is a matter of obtaining data and performing the appropriate calculations. Subsequent steps pertain to this.

Define the Criteria for Evaluating Alternatives

To quantify a decision, measurable criteria for evaluating the decision alternatives must be defined. Crucially, each decision alternative should be measured against the same criteria, and these criteria should measure what is ultimately of interest to the decision-maker.

In drug development the key criteria are commonly cost, time, and probability of the drug gaining marketing approval and reimbursement. In this example, these criteria are used. "Cost" is defined as the total development costs from the phase IIb study onwards (sunk costs are not included, as these can no longer be influenced by future decisions). "Time" is the number of months until the last patient last visit (LPLV) of the phase III study, which could be a different time in Japan from the rest of the world if a stand-alone Japanese phase III study is performed. Probability of marketing approval and reimbursement is assessed separately for Japan and the rest of the world.

It is generally unlikely that any one alternative will be best according to all three decision criteria. For example, an alternative that is more likely to succeed almost certainly will cost more and take longer. Cost, time and probability are also such disparate concepts that they may be hard to trade off in the mind of the decision-maker. To help decision-makers understand the trade-offs among these three criteria, a decision model that includes both development and market components is used to combine cost, time and probability of success into an overall measure of *expected* (i.e., probability-weighted) *net present value* (eNPV). The market model for the drug predicts the market size (number of patients) and share (percent of patients) over time in each region of the world. The longer it takes for the drug to be marketed the smaller the predicted revenue, and this revenue is discounted by both the probability of success and the rate of time preference for money. The cost of development is then subtracted from this discounted revenue to determine the eNPV.

A weakness of this approach is that it uses the market model for a purpose for which is it not intended. Such models are used primarily for making high-level portfolio decisions and cash flow predictions. They are typically based on historical analogues for similar drugs and order of market entry. In the market model used in this example, the predicted market share does not depend on drug efficacy, safety, price or strength of evidence, so is unlikely to be reliable for making absolute predations of eNPV. Despite this, a market model is used to distinguish between the alternatives as the relative differences between the alternatives are likely to be robust. Ultimately choosing the alternative with the highest eNPV is of interest, and errors in its prediction likely affect all alternatives in a similar way.

Assess Information and Uncertainty

Having defined the alternatives and established how we will choose between them, it is necessary to assess what information is needed to link the alternatives to the criteria. Some of this information could be assumed to be known exactly, for example the sample size of a study (which is actually a decision already made). But in general this information is uncertain, and it needs to be quantified and incorporated into the decision model.

Using Influence Diagrams

A useful tool to assess what information is needed is an *influence diagram*. This is a directed acyclic graph (DAG) comprising three types of nodes: decision, uncertainty, and value. These are conceptually the same as the decision, uncertainty and end nodes of the decision tree. A DAG is a formal mathematical description of a model and is used as a tool to structure the links between the nodes. The influence diagram for the case study is shown in Fig. 7.2.

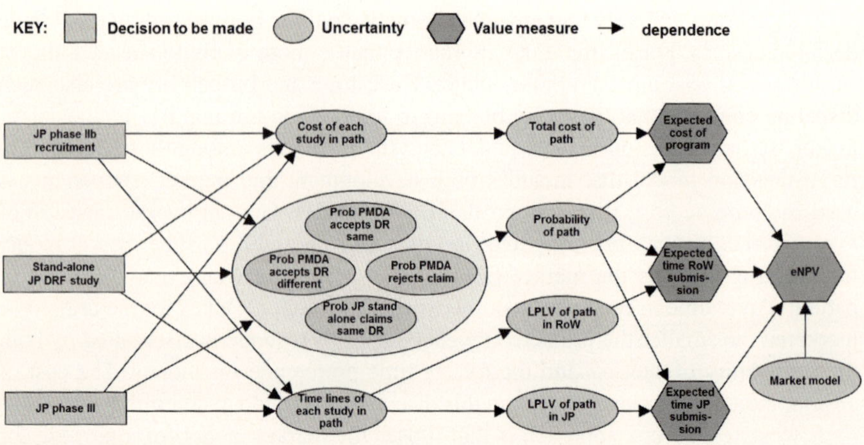

JP = Japan, RoW = Rest of world, LPLV = last patient last visit, eNPV = expected Net present value, DR = dose-response

Fig. 7.2 Influence diagram

Firstly, we denote value nodes by hexagons. These are the criteria that are used to evaluate the alternatives, and they should ideally be placed at the right side of the graph. The children of these nodes define what is needed to calculate the parent node, and the directed edges linking the two denote the dependence between them, which could be probabilistic or deterministic. In this case in order to find the eNPV, it is necessary to know the expected cost of the program, the expected times to submission in Japan and the rest of the world, and the expected revenue from the market model.

Secondly, the uncertainty nodes are denoted by ellipses. Knowledge of drug development is used to assess what is needed to calculate each of the children nodes from the children nodes produced by the previous step. The market model node is known, so this is seen as a basic piece of information and needs no further children. To find the expected cost of a development program, the cost of each possible path through the decision tree and the probability of each path need to be known. Similarly, the expected submission times are derived from the LPLV of the final study in a path and the probability of that path. The path probabilities are found from the product of all the probabilities on the path. To simplify the DAG we group these together in a node. The total costs comprise the costs of each study. LPLVs are found from the timelines of the studies. These uncertainty nodes, with no further uncertainty node children, define the basic information that is needed to populate the decision model; they include the cost and time to LPLV for each study, the probabilities of the PMDA accepting claims, and the probability that a Japanese stand-alone phase IIb study leads to the claim that there is a global MED. Sub-influence diagrams are drawn for the cost and time to LPLV of a study, which depend on the sample sizes, number of sites, recruitment rates, costs per patient, and time between studies. A list of the studies needed in the different alternative strategies is given in Table 7.3.

Finally, the decision nodes are denoted by rectangles, and they are linked to the uncertainties which they influence.

All the information that is needed to populate the decision model is described in Table 7.5. Some of this information is known exactly, e.g., the sample sizes and other design features of the studies, but most of it is uncertain. The costs for the studies have already been estimated for budget calculations. Similarly the market model for the drug has already been constructed, and outputs from this can be used. The recruitment rates can be estimated from looking at historical rates, by region, from studies in the same indication from an internal database. Uncertainty is expressed as a plausible range of values, defined by the 10th and 90th percentiles of a probability distribution over possible values. These ranges are elicited from expert opinion.

Using Conditional Probability

The probabilities of the PMDA accepting the claims are hard to elicit directly. These are broken down into probabilities that are easier to elicit, and the required probabilities are assembled using the rules of conditional probability.

Table 7.5 List of information needed for the decision model. Study information (design, recruitment, and costs) will be different for each study, and probabilities will differ depending on the choice of studies

Information type	Details of information needed
Study design	Total sample size
	Sample size needed in Japan
	Exposure time (weeks)
	Number of sites in the study by region (North America, Europe, Japan, and others)
Recruitment	Proportion of sites active (start recruiting patients) at the start of the study (by region)
	Time in weeks for all sites to become active (by region)
	Recruitment rate (number of patients per site per month)
Costs	Cost per patient by region
Probabilities	Probability that PMDA accepts the claim that the MED is the same in Japan and the rest of the world
	Probability that PMDA accepts the claim that the MED is different in Japan and the rest of the world
	Probability that the PMDA rejects claim
	Probability claim that MED from the Japan stand-alone study is the Global MED from phase IIb
	Probability claim that the MED in Japan is the same as the MED in all patients from a stand-alone Japanese dose range finding study
	Probability of technical and regulatory success given a drug is in phase IIb
Market model	Market size over time by region (number of patients)
	Patient share over time by region (proportion of patients)
	Price of drug per patient per day by region (adjusted for compliance)
	Time preference discount rate

For example, the probability that PMDA accepts the claim that the MED is the same in Japan and rest of world is the product of:

1. Probability that PMDA accepts the claim, given that this is claimed.
2. Probability that the phase IIb study claims MED is the same in Japan and the rest of the world, given this is truly the case.
3. Probability that MED truly is the same in Japan and the rest of the world.

A key principle of elicitation is to elicit quantities that the expert has an opinion on, and then use mathematical manipulations to map to the quantity that is needed for the decision model [3]. These three probabilities are easier to understand than their product, and different people have different expertise on each of these probabilities. Probability (3) above is elected from an expert with knowledge of the pharmacokinetics and pharmacodynamics of the drug. Probability (2) is a property of the power of the study. Probability (1) is elicited from an expert with experience of the PMDA.

Evaluate Alternatives Using Sound Models

Decision Model Computation

The decision model was coded using the statistical programming language R-2.15.1 [5] on a Linux computing cluster. The computations are intensive and involve simulating individual patients from several clinical studies under many different sets of assumptions and hence were run in parallel. The influence diagram helps to structure the code. Figure 7.3 gives a high level flowchart of the computer code and has a similar structure to the influence diagram.

Firstly, start with a decision alternative and a path though the decision tree. Read off the study design details on this path, along with the recruitment rates and costs. These are the inputs to the "study simulation" function, which returns the costs and LPLV in Japan and the rest of the world for each study on that path. The object-oriented nature of R is exploited when performing the study simulations. There is only one study simulation function. Its input parameters can describe any of the studies in the decision alternatives, and it returns the same data structure class for all of these studies. This simplifies the task of simulating the different sets of studies that make up each development program. Individual patient simulation is required, as it may happen that a global phase III study which includes Japanese patients is competing for patients with a Japanese stand-alone phase IIb study. To account for this, it is necessary to keep track of when each site in Japan is recruiting in both studies and adjust the recruitment rates accordingly.

Secondly, once the costs and times for each study in each path have been simulated, they and the market model data are taken as inputs to the "program simulation" function. This calculates, for each path, the total costs of all the studies in a development program, the LPLV time of the last study to finish, and the NPV.

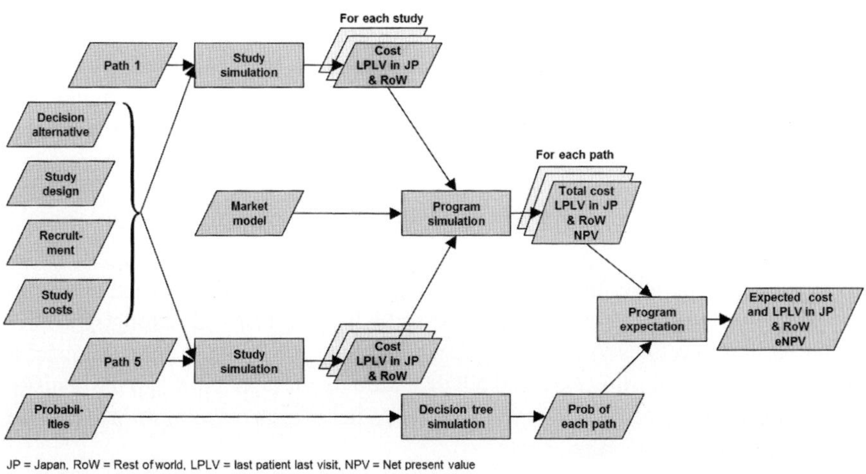

Fig. 7.3 Computer program flowchart

Thirdly, the probabilities of each path through the decision tree are calculated, and these, along with the outputs from the "program simulation," are used as inputs to the "program expectation" function, which returns the expected costs, times to LPLV in Japan and the RoW, and the eNPV over the decision tree paths.

Fourthly, this process is repeated for each set of decision alternatives.

Results

Table 7.6 shows the results of the decision model. For each of the six decision alternatives, the expected incremental costs, times, and NPV, relative to the ideal case, are shown. One wants an option with low extra costs, low extra time to LPLV, and the least loss to eNPV. The results show that the choice to extend the recruitment of the global phase IIb study in Japan is better than stopping the study when there are enough patients for the overall study analysis. This option costs less, is quicker in Japan and the rest of the world, and has the lowest eNPV loss. It is unusual for a decision alternative to dominate on all value measures. As this is the case here, no value trade-offs are needed.

For either choice of global phase IIb design, the choice of stand-alone Japanese studies makes very little difference in terms of cost or time to LPLV in the RoW, but an adaptive phase IIb/III stand-alone Japanese study leads to the shortest time to LPLV in Japan, so it is preferred.

DA is not a "black box," and performing it gives insights into why different decisions affect desired values. Not extending Japan recruitment for the phase IIb study will likely lead to stand-alone studies being needed, as the PMDA would most likely not accept that the MED in Japan is the same as in the rest of the world. These stand-alone studies are expensive and take a long time. Furthermore, not having Japanese patients in the global program will cause delays for the global submission studies, as recruitment will be slower without any sites in Japan. Conversely, having Japanese patients in the global program slows down recruitment for the Japanese stand-alone studies, as sites will be competing for the same patients. If the recruitment extension is done for the phase IIb study, it is likely that the stand-alone Japanese studies will not be needed, and the extra costs of this option are mostly due to the recruitment extension itself. The expected value of the delay is due primarily to the probability that a stand-alone program is needed.

The concept of expected (i.e., probability-weighted) costs, times, and NPVs can be difficult to communicate, as it is not necessarily a result that might ever be observed, but it is rather the probability-weighted average over all possible futures. It is easier to understand the costs, times, and NPVs for one path through the decision tree, as this relates to something that could happen. To aid the communication of the analysis, the costs, times, and NPVs for the likeliest path through the decision tree for each decision alternative were derived. These results are not shown, but they lead to the same decision as above.

Table 7.6 Expected outcomes for each decision alternative

Strategy		Expected value measures			
Phase IIb	Stand-alone JP studies IF needed	Costs (Mill USD, vs. ideal)	Time LPLV RoW (Months, vs. ideal)	Time LPLV Japan (Months, vs. ideal)	NPV (Mill USD vs. ideal)
Current—stop at 50 patients	3 arm JP P2b and join G-P3 at risk	26.1	0	48	−82
	Combined stand-alone P2b/3	26.0	2	18	−71
	2 arm JP P2b and stand-alone P3	25.0	2	52	−101
Extension—recruit to 120 patients	3 arm JP P2b and join G-P3 at risk	7.8	0	7	−21
	Combined stand-alone P2b/3	7.8	1	3	−20
	2 arm JP P2b and stand-alone P3	7.7	1	7	−23

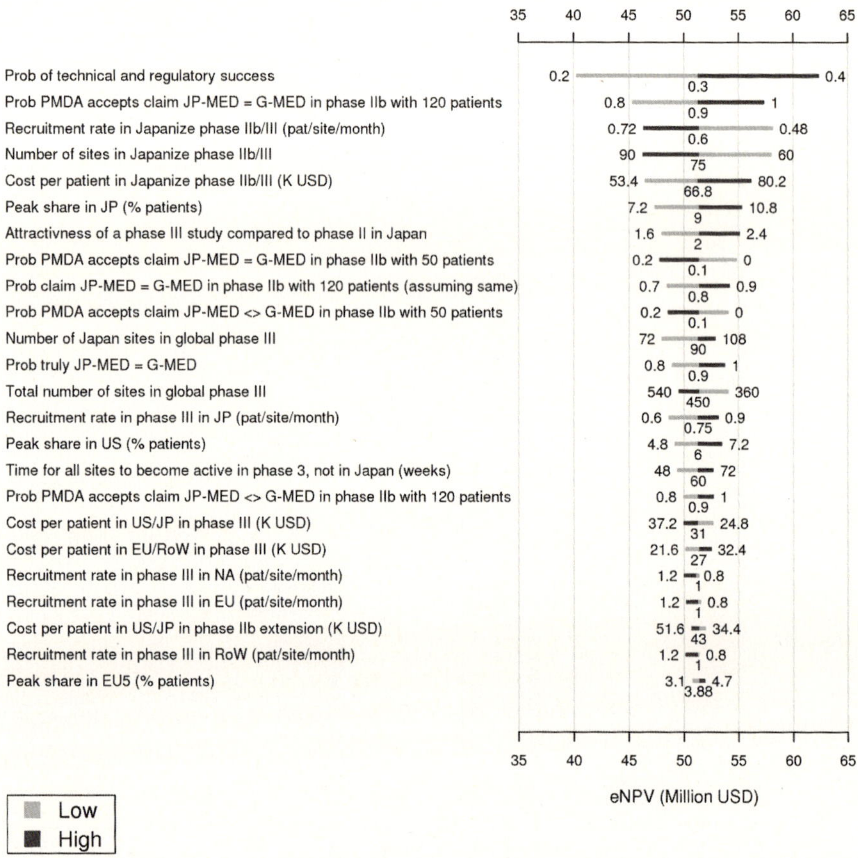

Fig. 7.4 Tornado plot of a one way deterministic sensitivity analysis, looking at the difference between the alternatives of (a) recruit 50 Japanese patients and (b) recruit to 120 Japanese patients in extension, both with a combined stand-alone P2b/3 study performed in Japan if needed

A key feature of a decision model is that it allows for uncertainty in the underlying assumptions to be represented and for the implications of that uncertainty to be assessed. A *one-way deterministic sensitivity analysis* on the model is performed, where each uncertain parameter in the model is changed, one at a time, to its upper and lower limits (10th and 90th percentile values), and the value (criteria) measures recalculated. For each possible parameter value, we calculate the incremental eNPV for (a) extending the recruitment of the global phase IIb study compared to (b) not extending the recruitment (with an adaptive phase IIb/III stand-alone Japanese study if needed). If this difference is positive then the recruitment extension is the preferred option, given the particular value of the uncertain parameter. The results of this sensitivity analysis are displayed in the *tornado diagram* in Fig. 7.4. The names of the uncertain parameters are shown down the left side, and each bar represents the effect on differential eNPV when that parameter is varied. The base case value for each parameter is shown under the bar, and the limits are given at each end of the bars. The bars are centered on the base case eNPV, and the ends of the bars are the eNPVs found at the parameter value limits. This shows us two important things. Firstly, the decision is most sensitive to parameters at the top of the plot, as these are the ones whose uncertainty has the most effect on the eNPV, and where it is most valuable to use resources to find more information. In this case the probability of technical and regulatory success is the most important uncertainty. Conversely, uncertainties such as the phase III recruitment rate have very little effect on the differential eNPV. Secondly, it is seen that in every case the differential eNPV is positive, so it is always best to extend the phase IIb study. This is stress-testing the robustness of the decision, and it can be used to identify situations that would lead to a different decision being preferred. A key principle of the value of information is that information only has value if it changes your decision. In this case, as reducing the uncertainty on any parameter would not change the decision, there is no value in collecting further information.

Stakeholders Commit to Implementation

A decision is not made until resources have been irreversibly committed. The final step is to gain the approval of the decision-maker to implement the recommendations from the DA. It is therefore vital that the decision-maker trusts the analysis. Two key ways to build this trust are to closely involve the decision-maker during each stage of the DA and to ensure clear communication of the analysis and final recommendations. A further way to build this trust is to repeatedly exposure the decision-maker to the DA paradigm over time.

In this case study the key recommendations were to:

- Extend the recruitment in Japan for the phase IIb study to ensure the recruitment of 120 patients in Japan.

Amongst the possible decision alternatives, this leads to the quickest time to submission in all markets. Even though Japan would join the global phase III study after it has started, this delays LPLV less than not joining at all. It is also the cheapest option because it reduces the probability of needing a Japanese stand-alone program. This means it has the highest eNPV (because it is quicker and cheaper).

- Plan for a phase IIb/III Japanese stand-alone study in case it is needed.

If the PMDA rejects the MED claim from the global phase IIb study, this adaptive design leads to the quickest time to submission in Japan and costs about the same as the other options, hence it has the highest eNPV.

The team accepted and acted on these recommendations.

References

1. Burman CF, Wiklund SJ (2011) Modelling and simulation in the pharmaceutical industry-some reflections. Pharm Stat 10:508–516
2. Matheson D, Matheson J (1998) The smart organization creating value through strategic R & D. Harvard Business School Press, Boston
3. O'Hagan A (2006) Uncertain judgements: eliciting experts' probabilities. Wiley, Chichester
4. PMDA (2007) Basic principles on global clinical trials. http://www.pmda.go.jp/english/service/pdf/notifications/PFSB-ELD-0928010.pdf
5. R Core Team (2014) R: a language and environment for statistical computing. http://www.R-project.org
6. Sheiner LB, Steimer JL (2000) Pharmacokinetic/pharmacodynamic modeling in drug development. Annu Rev Pharmacol Toxicol 40:67–95

Chapter 8
Indication Sequencing for a New Molecular Entity with Multiple Potential Oncology Indications

Jack Kloeber Jr., Alex Stojanovic, and C. Kwon Kim

Abbreviations

BC	Metastatic breast cancer
DPL	Decision programming language by syncopation
eNPV	Expected NPV, a probability weighted average of all possible NPV outcomes, including technical risk
G-BA	Gemeinsamer Bundesausschuss (Federal Joint Committee, the supreme decision-making body of the healthcare system in Germany)
IP	Intellectual property
IQWiG	Institut für Qualität und Wirtschaftlichkeit im Gesundheitswesen (scientific unit that assess effectiveness, quality, and efficiency of diagnostic and therapeutic methods and pharmaceuticals in Germany)
MM	Multiple myeloma
MODA	Multiple objective decision analysis
NPV	Net present value, a sum of discounted cash flows over a defined time horizon using an agreed upon discount rate
NSCLC	Non-small-cell lung cancer
PC	Prostate cancer

J. Kloeber Jr., Ph.D. (✉)
Kromite LLC, 148 East Street Road, Suite 121, Feasterville, PA 19053, USA
e-mail: jkloeber@kromite.com

A. Stojanovic, Ph.D.
Phosphagenics, 36 Fountayne Ln, Lawrence Township, NJ, USA
e-mail: astojanovic@phosphagenics.com

C.K. Kim, Ph.D.
Kromite LLC, #202-2446 Point Grey Road, Vancouver, BC, Canada V6K-1A2
e-mail: kkim@kromite.com

Z. Antonijevic (ed.), *Optimization of Pharmaceutical R&D Programs and Portfolios: Design and Investment Strategy*, DOI 10.1007/978-3-319-09075-7_8

PD	Pharmacodynamics
PK	Pharmacokinetics
POS	Probability of success (technical success for a phase)
PTD	Patient treatment day
PTRS	Probability of technical and regulatory success (combined POS for all phases)
PV	Present value
RA	Rheumatoid arthritis
R & D	Research and development

In this chapter, we address portfolio optimization at a sub-portfolio level. Even a "mini" portfolio—such as a single molecule with numerous indication opportunities—can pose a significant challenge to an organization; there are many interdependencies, correlations, and synergies to consider, weigh, and trade off. Here we present a case study that considers the challenge of optimizing this sub-portfolio by determining the best sequence of three potential indications. The same methods can be applied to a molecule with even more indication opportunities, a group of related molecules, or an entire portfolio.

Introduction

Every pharmaceutical or biotechnology company aims to maximize the value of a new molecule. One common strategy involves securing, early on, the intellectual property (IP) rights for a variety of potential indications in which there is a hint of effectiveness. But once a company has been granted a patent for multiple indications, what is the best development strategy for the molecule? Here are three interesting and real conundrums related to indication sequencing.

Situation 1: Avastin® is approved for use in Metastatic Colorectal Cancer (US: 2004; EU: 2005), Advanced Non-Small-Cell Lung Cancer (NSCLC; US: 2006; EU: 2007), Metastatic Kidney Cancer (US: 2009; EU: 2007), Glioblastoma (US: 2009), Metastatic Breast Cancer (US: 2008—Revoked in 2011; EU: 2007), and Advanced Ovarian Cancer (EU: 2011); tested in Unresectable Gastric Cancer, Nonmetastatic Colon Cancer, Pancreatic Cancer, Pediatric Osteosarcoma, Leiomyosarcoma, Age Related Macular Degeneration, and potentially others. A very successful product; however, were there opportunities for an even more efficient sequence? Were there hints earlier on in development that Metastatic Breast Cancer was a high-risk indication that should have been deprioritized?

Situation 2: Topamax® (topiramate) was approved to treat Epilepsy in adults and children (US: 1996), then Migraine (US: 2004) but eight years after the first indication and only two years prior to patent expiry; additionally, in 2010, Ortho-McNeil was fined $6.14 million by the FDA for promoting Topamax to treat Psychiatric Disorders without applying for any Federal government approval. In 2012, as a generic, topiramate in combination with phentermine received approval for Weight Loss. Was there an opportunity for another indication(s) that was not appropriately considered early on in development?

Situation 3: Remicade®, Humira®, Enbrel®, Orencia®, Stelara®, and Simponi® have all been approved as biologics for use in various inflammatory conditions (among other products). Remicade® was first approved in Crohn's Disease (followed by five other indications), Humira® was first approved in Rheumatoid Arthritis (RA; followed by six other indications), Enbrel® was first approved in RA (followed by four other indications), Orencia® was first approved in RA (studied in various other indications), Stelara® was first approved in Moderate to Severe Plaque Psoriasis (followed by Psoriatic Arthritis), and Simponi® was first approved in RA (followed by two other indications). Four of these six were initially approved in RA. Enbrel® was the first to receive approval in RA, but was that indication the right launch strategy for the other three? Was not launching first in RA the right strategy for Remicade® and Stelara®?

Whether the decision is about which indication should be (a) the focus in assessing efficacy and safety in the first exploratory Phase IIa study, (b) the first submitted for regulatory approval, or (c) the focus of initial negotiations regarding reimbursement and market access, each pharmaceutical company decides on the sequence for evaluating and launching into the potential indications. Additionally, many New Product Development Teams struggle with the following questions:

- How should we optimize the sequence of developing follow-on indications?
- Should we consider each indication as a separate project, independent of the other indications?
- What should we assume about launch date, wholesale price, effect on the sales force, probability of success?

- Is the complete value of the molecule simply the sum of the individual indications?

Using an Oncology case study, we illustrate and compare three different approaches for choosing the sequence of indications to pursue for a molecule's development program:

1. Simple (Indication) Ranking Method
2. Decision Tree Method (analysis at the molecule level)
3. Multiple Objective Decision Analysis (MODA)

Case Study Background

A potent new Oncology drug candidate (BRT104) has shown tremendous promise in preclinical studies and is currently being assessed within a Phase I first-in-human trial. The company has a significant amount of experience in oncology drug development, a strong field force, and an existing portfolio of products in Oncology. Based on a multidisciplinary team analysis, three Oncology indications are judged to be scientifically, medically, and commercially most promising: Multiple Myeloma (MM), Metastatic Breast Cancer (Stage IIIc/IV; BC), and Prostate Cancer (PC).

Management has made a policy decision (a decision that is not to be questioned) that only one indication will be developed at a time for BRT104, in order to mitigate the R & D cost impact in any one year and reduce the risk associated with investigating a new mechanism of action (i.e., only one Phase II trial will start in any given year). Therefore, the second indication would launch 1 year later, and the third indication would launch two years later.

Simple Ranking Method

The most common decision making method involves a product development team selecting the initial indication by simply ranking each indication on selected key measures; these measures should be aligned with the company's objectives. This method can be quite simple, relying on as little as one measure (e.g., Net Present Value; NPV) or slightly more complex. For this case, as shown in Fig. 8.1, seven distinct measures were developed, each of which supports one of the top level company objectives of Strategic Fit, Technical Risk, Commercial Potential, or Financial Health. The New Product Development Team assessed these measures for each of the three indications being considered.

Figure 8.1 reflects an example of how the Simple Ranking Method can be applied. To conduct an indication sequencing analysis, using any of the three methods, requires a significant amount of research, market understanding, and clinical foresight. A wide variety of sources, both internal and external, should be utilized to collect data relevant to each indication with respect to the desired metrics. For our

	Strategic Fit	Technical Risk		Commercial Potential		Financial Health	
Indication	Degree of Unmet Medical Need *	Risk (PTRS)	Remaining R&D Cost ($M)	Market Size (Patients, US)	Peak Sales ($M)	NPV ($M)	Risk Adjusted NPV ($M)
Multiple Myeloma	6	25%	~ $200M	~ 22,500	~ $540M	~ $600M	~ $130M
Metastatic Breast Cancer	8	15%	~ $150M	~ 45,000	~ $1,000M	~ $1,400M	~ $190M
Prostate Cancer	2	7.5%	~ $250M	~ 195,000	~ $1,700M	~ $2,500M	~ $175M

Fig. 8.1 Simple ranking method. List of measures and the assessment/data for the three indications: *green* indicates highest scoring, *yellow* is second highest, and *red* is lowest; *scale of 1 (low) to 10 (high). Degree of Unmet Medical Need was based on 5-Year Survival Rates and other internal expert assessments. Remaining R & D Cost assumes Phase II/III studies for registration, additional preclinical studies, and cost of regulatory filing, but no additional Phase IIIb/IV studies for Market Access (shown as NPV in $M); discounted value shown. Data was collected through a variety of internal and external sources including leading cancer Web sites [1–4]

case study, the National Cancer Institute and American Cancer Society Web sites, among others, were leveraged to determine the Degree of Unmet Medical Need (based on 5-Year Survival Rates) and Market Size (Patients, US). While data driven measures are desirable, subjective assessment by approved experts may be needed. There is significant value in assessing these key measures, but the correct choice for the team is not at all clear.

Should the company choose to develop BRT104 for Multiple Myeloma because the molecule has the *best chance of launching* in that particular indication (i.e., highest assessed PTRS—Probability of Technical and Regulatory Success), for Metastatic Breast Cancer because it offers BRT104 the greatest opportunity to align with the company's objective of *addressing high unmet medical need* (while serendipitously requiring the lowest upfront investment for R & D), or for Prostate Cancer because it is the indication that provides by far the *highest peak sales*, *potential NPV*, *and even risk-adjusted NPV*? Clearly, the use of a Simple Ranking Method presents a lot of information and insight, but does not effectively support the decision and presents the team with certain limitations (see below).

Simple Ranking Method: Open Questions

- Is the company's objective (a) to select the best *indication*, or (b) to maximize the value of the *compound* through an optimized sequencing of indications?
- If the "Best" indication is selected, what assumptions should be made regarding the sequencing of follow-on indications and the overall value of the molecule?
- When one indication scores highly on some measures and other indications score highly on other measures, which measure(s) is most appropriate to utilize for selecting an indication?

- If point estimates are used, how valuable are the estimates, given the early stage of the molecule (e.g., how confident are we in an NPV estimate when BRT104 has completed only Phase I)?

Decision Tree Method

A second, more holistic decision making approach, addresses some of the questions raised in the Simple Ranking Method; it leverages decision analysis principles to improve decision quality and structure the problem using a *Strategy Table*, and an *Influence Diagram/Decision Tree*.

The focus shifts from evaluating the development strategy for an initial indication (the limited focus of the Simple Ranking Method) to the three-indication launch and lifecycle management strategy. Such a method generates much greater insights and helps the team develop a coherent long-term product strategy. However, it also adds complexity to the decision. In this example, the assessment and decision are made purely on the basis of expected NPV of the molecule, although other objectives and associated measures could be leveraged.

For our Case Study, a Strategy Table was developed (Table 8.1) to outline three coherent indication sequencing strategies (i.e., molecule development strategies) for BRT104: Cheapest First (Metastatic Breast Cancer first), Easiest First (Multiple Myeloma first), and Biggest First (Prostate Cancer first). The strategies were developed by the project team to be differentiated, and simple names were used to describe the major drivers of the strategies.

Using the Strategy Table, the team could include interdependencies between the three indications and make some molecule level assumptions for BRT104:

- Pricing of the first indication sets pricing for all subsequent indications (note: in a more advanced version of the analysis, a fourth strategy could also include additional health economic studies to justify a higher price for a second or third indication, even if a lower price was given initially for the first indication).

Table 8.1 Decision tree method: strategy table

	Elements of strategy			Strategy level data		
Strategy	First indication	Second indication	Third indication	Year of first launch	PTRS of first indication (%)	Pricing ($/PTD)
Cheapest first (low cost)	BC	MM	PC	2018	15	350
Easiest first (best chance)	MM	BC	PC	2019	25	450
Biggest first (go for the gusto)	PC	BC	MM	2018	7.5	250

Three strategies are shown (each strategy = indication sequence), along with examples of key data for assessing the strategies; MM = multiple myeloma, BC = metastatic breast cancer, PC = prostate cancer, PTD = patient treatment day

- Development of the other indications would continue regardless of success or failure of the first indication (note: in a more advanced version of the analysis, the PTRS of subsequent indications could be modified based on the success or failure of the first indication if the team feels that there is some interdependency; for example, there might be a common safety risk, PK or PD properties that are inadequate for any of the indications).
- Regardless of sequencing, the patent for BRT104 would expire in 2028; therefore, pursuing the Easiest First strategy, with a 2019 Launch Date, would result in one less year on the market.

What's important to recognize is that the Decision Tree Method combined with generation of a Strategy Table already helped the team think about BRT104 objectives at a molecule level, something that was missing during the utilization of the Simple Ranking Method. In this example, we chose to maximize the overall NPV of the molecule. Following the creation of an aligned Strategy Table, the team moved to develop an Influence Diagram (Fig. 8.2). The Influence Diagram shows a high-level relationship of the three strategies and the associated values and uncertainties. It also forms a straightforward schematic for the development of a more powerful decision analysis model.

The base values and associated level of uncertainty for each of the key variables were then determined. Uncertainty was incorporated using the 10th (low), 50th (base), and 90th (high) percentile values. For this Case Study, the key variables included the *Probability of Success* (*POS*) of the Phase IIa trial, the Pivotal Trials, and the Regulatory submission, giving an overall *Probability of Technical* and *Regulatory Success* (*PTRS*). This also allowed for the consideration of scenarios where investment in R & D is made without any return (e.g., indication fails in Pivotal

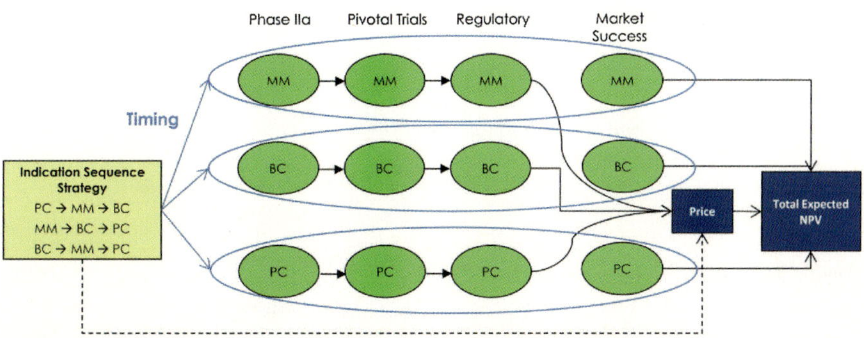

Fig. 8.2 Decision tree method: influence diagram. The key variables (technical risk, market success, price) and their relationship to the three strategies to be evaluated; alternatively, what influences Total Expected NPV, what influences those factors, and how selecting one of the three strategies influences Total Expected NPV. *Green ovals* represent outcomes with an associated uncertainty (e.g., the uncertainty of a positive outcome of a Phase IIa Trial in MM), while the *blue boxes* represent numerical values or calculations (e.g., Price); MM = multiple myeloma, BC = metastatic breast cancer, PC = prostate cancer. DPL 7.0 (Decision Programming Language 7.0, Syncopation Software, Concord, MA) was used to model the Influence Diagram

Indication	Probability of Success			Total PTRS
	Phase IIa	Pivotal Trials	Regulatory	
Multiple Myeloma	50%	55%	90%	25%
Metastatic Breast Cancer	35%	50%	85%	15%
Prostate Cancer	25%	40%	75%	7.5%

Indication	Market Success (PeakVolume, M PTD) 10th- 50th- 90th Percentiles		
	Launch Now	1 Yr Launch Delay	2 Yr Launch Delay
Multiple Myeloma	0.8-1.2-1.6	0.6-0.9-1.2	0.5-0.8-1.1
Metastatic Breast Cancer	2.2-2.8-3.4	1.6-2.1-2.6	1.5-1.8-2.2
Prostate Cancer	4.6-6.8-9.0	3.5-5.1-6.8	3.0-4.5-6.0

Fig. 8.3 Decision tree method: technical risk and commercial uncertainty. (**a**) (*Left*) Probabilities of Success (POS) shown for each stage of the development pathway, for each indication, along with Total PTRS (Probability of Technical and Regulatory Success). Success in any one stage is assumed to mean achieving endpoints which would support the Target Product Profile; (**b**) (*Right*). Values at the 10th, 50th, and 90th percentile for each Indication shown for Market Success (Total Volume). Values are shown for Base Case launch timing, one year delay, and two year delay, which depend upon the sequencing. Select variables are shown for illustration

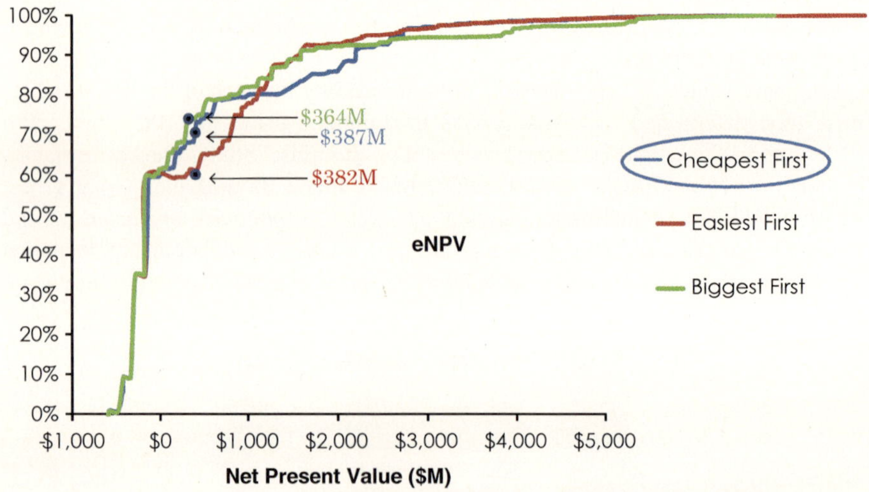

Fig. 8.4 Decision tree method: analysis. Cumulative distribution of the eNPV values of the three strategies

Trials and never reaches Regulatory Filing). In addition, commercial variables included an overall measure of Market Success (Peak Volume) and associated Price. Figure 8.3a, b shows examples of data used to assess the variables in this Case Study.

Once all of the values and associated uncertainties were assessed, we incorporated them into the decision tree model. In this particular Case, the Indication Sequence Strategy with the greatest (eNPV) would give an expected NPV of $387M (eNPV of *$387M for BC-MM-PC* (*Cheapest First*), eNPV of $382M for MM-BC-PC (*Easiest First*), and eNPV of $364M for PC-BC-MM (Biggest First)).

Launching BRT104 first in Metastatic Breast Cancer followed by Multiple Myeloma and, Prostate Cancer, yields the highest expected NPV for BRT104. While these figures represent the eNPV of each strategy, the distributions around these averages represent the hundreds or thousands of possible scenarios that may surface from the sequencing decision and the associated uncertainties following that decision.

For example, Fig. 8.4 shows that no one strategy is completely dominant over the other two strategies. The *Cheapest First* strategy does have the slightly highest expected NPV and it has the highest chance of generating a $2B return (16 % compared to 8 % for the other two strategies. However, the *Biggest First* strategy has about a 7 % chance of generating more than $3B in NPV, while the other two strategies have a less than 1 % chance of doing so. Hence, the "Go for the Gusto" motto was presented for this strategy. Therefore there is no dominating strategy of the three being analyzed.

Decision Tree Method: Limitations

The Decision Tree method addresses many of the shortcomings of the Simple Ranking Method, but still leaves several unresolved issues:

- Commercial value of indications at significantly early prior to proof of concept and more than 5 years out are difficult to estimate and involve a high degree of uncertainty. In our opinion, a $5M difference in eNPV is immaterial, and, in this case, should not drive an important decision.
- A Decision is being made based solely on one financial objective (in this case Long Term Value) ignoring other *nonfinancial* objectives such as strategic fit, unmet medical need, or other *financial* objectives such as long term revenue growth (CAGR). The decision maker is forced to "guess" about a strategy's alignment with these non-assessed objectives.

Multiple Objective Decision Analysis (MODA) Method

The most comprehensive approach available to pharmaceutical companies today is *Multiple Objective Decision Analysis* (*MODA*) [5, 6]. MODA integrates multiple, often competing objectives, when assessing the value of a compound and different strategies for that compound. These objectives can be financial in nature, as discussed above (e.g., eNPV or top-line revenue); in a growing number of earlier stage pharmaceutical situations, they are largely nonfinancial (e.g., align with corporate strategy, maximize patient centricity, address the greatest unmet medical need). The MODA Method involves seven key steps, which allow for the multi-objective assessment and comparison of different strategies:

1. Identifying the desired objectives
2. Structuring objectives into a hierarchy
3. Developing a measure and value function for each objective

4. Assigning weights or relative importance to objectives
5. Developing creative molecule strategies
6. Assessing each strategy on each objective
7. Including technical risk and investment
8. Conducting an analysis to identify the best strategy (or produce a better hybrid strategy)

Often, teams responsible for molecule development are multidisciplinary and global in nature, adding to the inherent complexity of decision making. A well-facilitated workshop(s) focused on elicitation of corporate and disease area objectives helps stimulate discussion and align the organization and key stakeholders on achievable and measurable objectives that are relevant to the compound.

In this Case Study, a multidisciplinary team identified objectives for BRT104 through focused facilitated workshops (Step 1). While Commercial Success was certainly the main objective, there were others of significant importance. The company intended to grow its presence in Europe, and also to ensure Strategic Fit with the current product portfolio and corporate vision. A team of functional experts (e.g., pharmacologists, clinicians, marketing experts) assisted in developing suitable measures aligned with each overarching objective, using molecule level information (as opposed to only indication level information). For example, the team determined that Strategic Fit of BRT104 could be measured by assessing how much each development strategy Emphasized Primary Care and Maximized Leadership in Oncology. Objectives, supporting objectives, and measures were then organized using an *Objective Hierarchy* (Step 2) to facilitate visual clarity of objectives. In some cases, multiple but nonoverlapping measures of a particular overarching objective were needed due to the breadth of the objective. Figure 8.5 shows the final set of objectives and measures.

A critical and often difficult part of this phase is identifying an adequate measure of a qualitative objective. A *Value Function* shows the relative value of achieving one level of a particular objective/measure (here, European Presence). For example, Fig. 8.5 shows how achieving Market Access in 60 % of Europe (measured as the % of total PTDs in Europe) would provide a value of 7.5 (scale from 0 to 10), while achieving 30 % might only provide a value of 2. In this fashion, the value function reflects the objectives of the company. It also provides insight about the key inflection points in the development or commercialization of the product; for example, this value function shown below suggests that a 30 % drop in EU Market Access—which is approximately the equivalent of losing Germany or the UK—would have a much more detrimental impact on the value to the company if that drop occurs from 60 to 30 % (a decrease of 6 on the value scale) than a drop from 90 to 60 % (a decrease of 2 on the value scale). Such insight generates additional multidisciplinary discussion and brings forth alternate strategies to be further evaluated. For example, one might foresee a strategy that incorporates one or two additional Phase II studies to satisfy the requirements of G-BA/IQWiG and secure German market access—especially if it appears that without the study, the access would only be at the 30 % level.

Some value functions were less complicated than that for % Europe with Market Access by 2020. Figure 8.6 shows two additional examples of Value Functions in addition to the one shown in Fig. 8.5; this includes a linear function and a stepwise

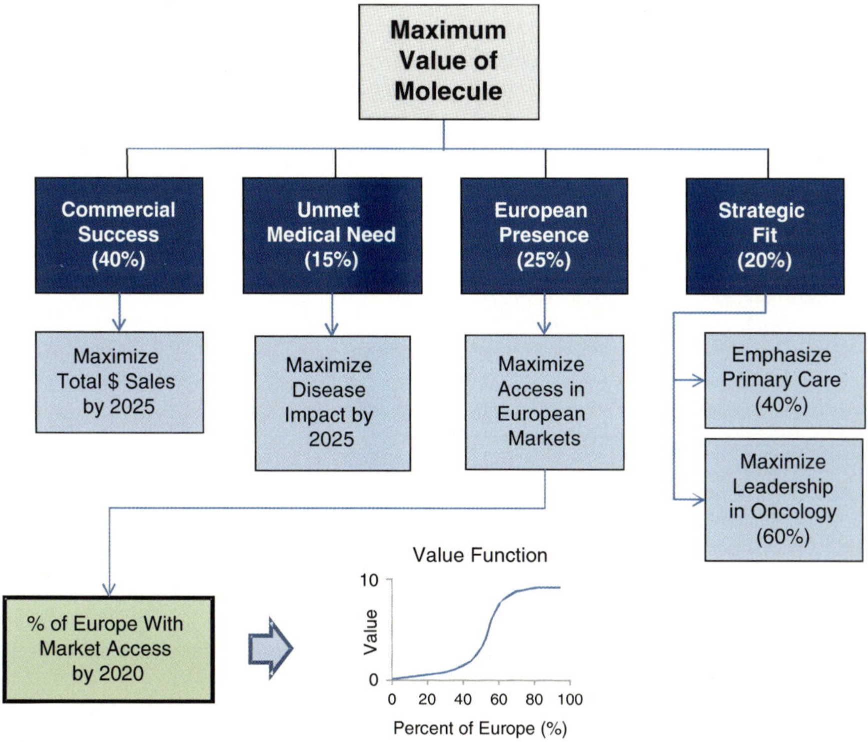

Fig. 8.5 MODA method: objective measurement and weighting. Management agreed-upon objectives, along with the associated weights for each objective (in *parentheses*). The breakout shows a Value Function for one of the objectives and measures—Percent of Europe with Market Access by 2020

or ladder function with only three levels for scoring. In all cases, the measure must be very well defined to avoid subjective scoring and to improve the communications achieved through these measures. For example, the measure *Improvement and Innovation over Standard of Care* has three categories. Each category must be quite well described so that the team members and management agree on what *Moderately Increase* means and how it differs from *Significantly Increase*. The category discussions also help drive the value assessment for achieving each of the categories.

Next, the Management team considered the relative importance of one objective over the other, assigning weights (Step 4; the relative contribution of each objective to the overall value). Figure 8.5 also shows the result of a second workshop which was focused on Steps 3 and 4 of the MODA Method. It is important to delay the weighting conversation with management until the measures have been completely agreed upon. The discussion of measures provides a definition of what is meant by each objective. Without such definition, stakeholders can easily talk past each other regarding what they think is meant by each objective. In some cases, having the measures in front of them dramatically changes their views of relative importance.

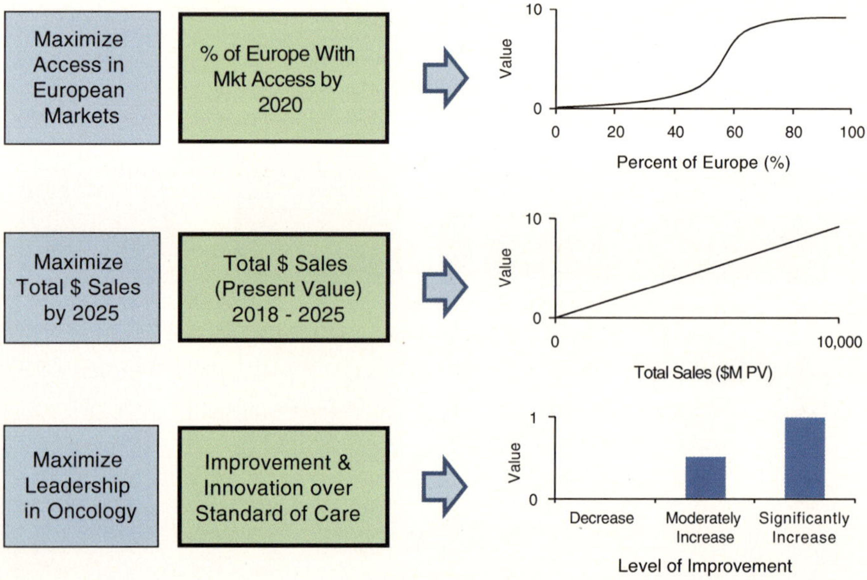

Fig. 8.6 MODA method: objective measurement and weighting. Three types of Value Functions are shown for three of the objectives and measures described in Fig. 8.5. Maximize Market Access in European Markets: *Curve*, Maximize Total $ Sales by 2025: *Linear*, Maximize Leadership in Oncology: *Ladder*

Table 8.2 MODA method: strategy assessment

Objective→	Commercial success	Unmet medical need	European presence	Strategic fit	
Measure→	Total $ sales (present value) by 2025[a]	Disease impact	% Market access in EU	Primary care emphasis	Leadership in oncology
Cheapest first	~$5,700M	13.5/30	20	None	Significantly increase
Easiest first	~$6,600M	13/30	60	Medium	Moderately increase
Biggest first	~$5,700M	11/30	40	High	Maintain

Three strategies are shown (Strategy = Indication Sequence), along with the assessment of each strategy along the six objective measures for four objectives

[a]Assumes that all three indications succeed; is not risk adjusted

In our example, Management determined that the most important objective for BRT104 was Commercial Success (40 % weighting), followed by European Presence (25 %). This discussion had to include a comparison of the range of PV of Total Success by 2025 vs. the range of % European Presence.

By incorporating this functionality, the MODA Method is considerably more comprehensive than the example Decision Tree Method, which is completely based on Long Term Financial Value (eNPV). Here, a strategy could provide the slightly lower NPV and still become the "Best" due to its alignment with other objectives.

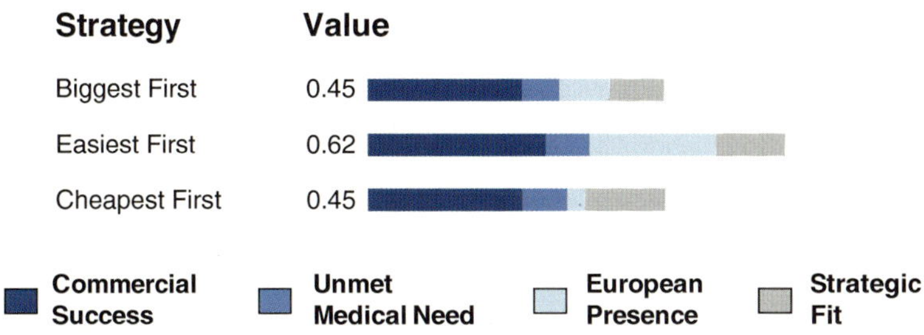

Fig. 8.7 MODA method: valuation. Three strategies are shown with the cumulative value according to the sum of the values of each objective (see Table 8.2 for values of each measure). The length of the bar shows the level of value contribution from each objective (standardized to a value between 0 and 1)

Each strategy was assessed on its ability to achieve each objective by objectively comparing the strategy characteristic to the measures (Step 5). Table 8.2 shows the data for each measure. Note that some of these strategy characteristics are arrived through

1. Simple summing of indication level characteristics
2. Assessing the lowest, highest, or average of all indication performances, depending upon the measure, or
3. Calculating a complex combination of indication characteristics (such as the decision tree analysis)

Experts in the Therapeutic Area are needed to ensure the competition aspect, future market demand, and other TA specific assumptions are part of the assessment. Once the Strategy-level characteristics are known, estimated, or calculated, appropriate values were easily determined using value functions shown in Figs. 8.5 and 8.6.

Before including technical risk into the assessment and analysis, we calculated the overall non-risk-adjusted MODA value for each strategy. This value is a combination of several financial and nonfinancial measures. Figure 8.7 shows the cumulative valuation of the three strategies under consideration. Interestingly, the inclusion of nonfinancial objectives and their associated measures resulted in the recommendation of a different strategy compared to the Decision Tree Method, as well as a richer and more insightful discussion.

The desire to address *European Presence* clearly drove the differentiation of the *Easiest First* strategy (starting with Multiple Myeloma). The team felt that BRT104 had a very strong case for high market access in Multiple Myeloma, but less so in the other two indications. At this point, it would be appropriate to discuss this impact of *European Presence* with the Project Team, possibly conducting a sensitivity analysis of the weight assigned this Objective.

As with all R & D assessment and decision making, omitting the impact of technical risk is unrealistic and misleading. To accomplish Step 7, the MODA assessment was automated to allow for immediate assessment of all potential outcomes for each strategy. In other words, each strategy could have the following outcomes.

Outcome Type 1: All indications succeed (in the order shown for each strategy).

Outcome Type 2: All indications fail.

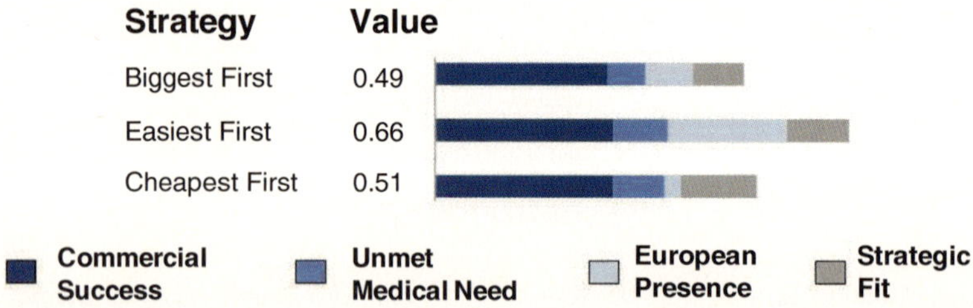

Fig. 8.8 Risk-adjusted MODA method: valuation. Three strategies are shown with the cumulative expected value according to the sum of the expected values of each objective (see Table 8.2 for values of each measure). The length of the bar shows the level of value contribution from each objective (standardized to a value between 0 and 1)

Outcome Type 3: Two indications succeed, one indication fails (this could occur three different ways since we have three different indications).

Outcome Type 4: Only one indication succeeds, the other two fail (this could also occur three different ways for each strategy).

Therefore, each strategy could have eight technical outcomes, assuming Success or Fail are the only outcomes for each strategy. The probabilities for each outcome were calculated and applied to the MODA values. MODA values have to be recalculated—e.g., if one indication fails while the other two succeed for a strategy, the MODA value should be lower since only two indications reach the patients and MODA values for each objective may change.

After applying all probabilities to all outcomes, and recalculating the many MODA values (automatically), we then calculated the overall expected MODA for each strategy giving us the chart shown below in Fig. 8.8.

This result is in contrast to a recommendation of the *Cheapest First* strategy when using a risk-adjusted financial measure alone (i.e., Decision Tree Method), which, although drives the highest risk-adjusted NPV, sacrifices Market Access in Europe. We see in this chart a simple comparison of three strategies with quite complicated underlying calculations. While the computations may be complex, the focus is always the same; use agreed upon measures and preferences to assess molecule-level characteristics of the strategies, and find the highest expected value strategy.

Step 7 also requires taking the development cost into account. Of course development cost for an indication depends upon the development success of each indication. For instance, if an indication is successful through Phase II, but fails in Phase III, the development cost must include Phase I, Phase II, as well as Phase III costs. Another strategy which fails in Phase II, has development costs of Phase I and Phase II. Taking this costing assumption into account, we calculated the expected cost for each strategy. The following chart, Fig. 8.9, compares each strategy's expected development cost vs. its expected MODA value. The decision makers can now decide on which strategy they should place their bets. Note that the three

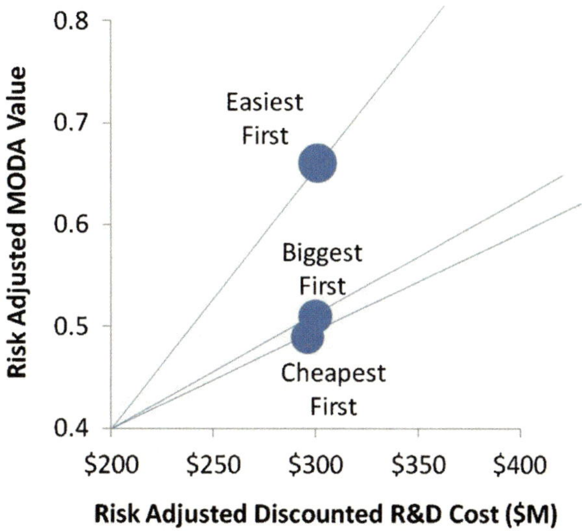

Fig. 8.9 Risk-adjust MODA vs. risk-adjusted discounted R & D cost. The steeper the slope of the lines drawn through each of the three strategy points, the more value the company should realize for each development dollar

expected costs are quite close. This is because the company's policy was to continue with all three indications, regardless of the outcome of each individual indication. If the policy changed to one that we have seen in action, there could be quite a different set of expected costs. An example of a simple policy is: If the first two indications fail, do not continue development for a third indication. A more sophisticated policy would base continuation of the following indications on whether this issue was safety or efficacy, and further, whether the issue appears to be a class, molecule, or disease related reason.

In Fig. 8.9, it should be clear that the *Easiest First* strategy point produces the steepest slope and thus produces the most expected value per development dollar.

Value of the MODA Method

In this Case Study, the MODA Method addressed several shortcomings of the Decision Tree Method. The consistency in value assessments and its transparency led to better and more open discussions about tradeoffs between strategies. Management particularly appreciated this facet of MODA, which led to better alignment around objectives and selecting a strategy that addressed a combination of objectives. Note the importance of Step 7 in accounting for risk and cost appropriately. As a general rule, costs should not be part of the Objectives Hierarchy and Risk, whenever possible, should be quantified and used appropriately in assessing the overall value.

Table 8.3 Assessment of three analytical approaches

Issue	Simple ranking	Decision tree	MODA
Indication sequence	X	X	X
Incorporation of uncertainties	X	X	X
Most appropriate measures		X	X
Indication dependencies		X	X
Molecule level analysis		X	X
Uncertainty of commercial value (early stage)		X	X
Incorporation of nonfinancial metrics			X

Comparison of the three approaches and the issues that each can successfully deal with as per the case study

Summary

In this Case Study, three different analytical approaches were evaluated for their value in helping to find the best indication sequencing strategy for a novel cancer therapeutic, BRT104. All of this modeling and interaction with a multidisciplinary team of experts and Management enabled the company to look at BRT104 in a new light. Clarity for the decision makers increased significantly and confidence that launching BRt104 first in Multiple Myeloma was the best way to meet multiple objectives was achieved for many stakeholders. Management gained trust in the working team and became confident that advancing BRT104 first in a Multiple Myeloma study captured the key objectives and main areas of uncertainty/risk.

This Case Study illustrates the particularly strong value of MODA, which addresses several more issues than does either a Simple Ranking Method or a financially driven Decision Tree Method (Table 8.3). MODA can be leveraged for other issues as well, including decisions regarding which projects to fund and other portfolio level decisions [7–9].

About KROMITE

KROMITE is a leading strategic advisory firm that specializes in the application of decision science to help clients make strategic decisions, manage risk, and create value. KROMITE was founded in 2003 to provide independent and unbiased support for tough decisions in the life science industry.

Our team, headquartered in New Jersey and located throughout North America and Europe, possesses unparalleled expertise in decision analysis and decision making. From years of working for pharmaceutical, biotech, medical device, and agricultural companies, our team commands intimate knowledge of tools, terminologies, organizational roles and responsibilities, R & D processes, common deal term structures, and organizational decision making processes, which allows our clients to rely on us as a partner and external expert.

For more information about KROMITE, please call us at +1 (267) 983 6305, email us at info@KROMITE.com, or visit our Web site at www.KROMITE.com.

References

1. National Cancer Institute (2013) http://www.cancer.gov/cancertopics
2. American Cancer Society (2013) Breast cancer facts & figures 2013–2014. American Cancer Society, Inc., Atlanta
3. Wikipedia (2013) http://www.wikipedia.org
4. Centers for Disease Control and Prevention (2013) National Program of Cancer Registries (NPCR). http://apps.nccd.cdc.gov/uscs
5. Keeney R (1992) Value focused thinking: a path to creative decision-making. Harvard University Press, Cambridge
6. Keeney R, Raiffa H (1993) Decisions with multiple objectives. Cambridge University Press, Cambridge
7. Kloeber J (2010) Deciding which projects to fund: using value focused thinking in pharmaceutical decision making. KROMITE LLC, Lambertville, White Paper
8. Kloeber J (2011) Current and cutting edge methods of portfolio decision analysis in pharmaceutical R&D. In: Portfolio Decision Analysis, international series in operations research & management science, part 3. Springer, New York, 162: 283–331
9. Kloeber J, Kim CK (2012) Horses for courses: portfolio management methods for different stages of drug development. KROMITE, LLC, Lambertville, White Paper

Chapter 9
Maximizing Return on Investment in Phase II Proof-of-Concept Trials

Cong Chen, Robert A. Beckman, and Linda Z. Sun

Introduction

The phenomenal expansion of our knowledge in molecular biology has led to an unprecedented number of new drug targets. However, the success rate for investigational drug candidates remains low while the cost for conducting Phase III confirmatory trials has risen rapidly. This makes Phase II POC trials more important than ever in late-stage drug development. In this chapter, we will investigate optimal cost-effective design strategies for randomized placebo-controlled POC trials. POC trials can also be non-randomized or use historical data as the control. However, the true type I/II error rates are hard to quantify. Although the methodology developed in this chapter is equally applicable to such trials, we do not support their routine application for POC purposes.

In the literature, there are two quantitative approaches to finding the optimal balance between benefit and cost. The first approach is to find optimal design parameters that minimize patient exposure at fixed type I/II error rates, e.g., under null as in

C. Chen, Ph.D. (✉) • L.Z. Sun, Ph.D.
Biostatistics and Research Decision Sciences, Merck Research Laboratories (MRL), UG1C-46, 351 N. Sumneytown Pike, Upper Gwynedd, PA 19454, USA
e-mail: cong_chen@merck.com

R.A. Beckman, M.D.
Center for Evolution and Cancer, Helen Diller Family Cancer Center, University of California at San Francisco, San Francisco, CA, USA

Oncology Drug Development, Daiichi Sankyo Pharmaceutical Development, Edison, NJ, USA

Z. Antonijevic (ed.), *Optimization of Pharmaceutical R&D Programs and Portfolios: Design and Investment Strategy*, DOI 10.1007/978-3-319-09075-7_9

Simon [14] or under any prior distribution for treatment effect as in Anderson [1]. This approach (hereafter referred to as sample size minimization approach) is appealing to statisticians because it is parsimonious and avoids assumptions such as the overall benefit of the study drug that could be controversial. As a result, numerous publications have been generated in the statistical literature. However, this approach has limitations when the choice of type I/II error rates itself is an issue and when the benefit of the study drug has to be taken into account. The second approach applies Bayesian decision analysis techniques to find the optimal design parameters by directly maximizing the net return (i.e., benefit–cost). It is used for determination of optimal sample size for Phase III trials subject to budget constraints [12] as well as for determination of Phase II sample sizes ([16] and [17]). Relevant work can also be found in Halpern et al. [9], Stallard et al. [18], and O'Hagan and Stevens [11]. This approach is appropriate when benefit can be quantified upfront and the parameter space for decision-making is very well defined. When benefit is overestimated, which occurs often in practice, such analyses tend to recommend a low bar for a Go decision to next stage of development, making it hardly acceptable to stakeholders [10].

We have proposed a new approach [4–6]. The idea is to find optimal cost-effective parameters by maximizing a benefit–cost ratio function. The numerator of the function is the expected number of truly active drugs correctly identified for Phase III development, each multiplied by the benefit per drug if applicable, and the denominator is the expected total Phase II/III development cost, including that resulting from Type I error. From a high-level perspective, the sample size minimization approach is equivalent to the use of our denominator as a utility function while assuming a constant numerator. The Bayesian decision-theoretic approach is equivalent to the use of the difference between our numerator and denominator as a utility function. One major difference among the three approaches resides on the way the intrinsic benefit of a study drug is handled. Our approach acknowledges the fact that variations in benefit may be important, and therefore incorporates them into the utility function in contrast to the sample size minimization approach which simply attempts to minimize cost. As compared to the decision-theoretic approach, our approach is less sensitive to small errors in estimation of benefit (and cost), which is notoriously difficult to estimate at the planning stage of POC trials. When only one trial is considered, the optimal design is independent of the benefit; when more than one trial are considered, the optimal designs depend only on the relative benefit which is considerably easier to assess than the absolute benefits that the decision-theoretic approach relies on.

The chapter is organized as follows. We describe the key design parameters in POC trial designs in section "Key Design Parameters". Section "Optimal General Strategy for Single POC Trials" introduces the basic form of benefit–cost ratio and applies it to derive the high-level strategy for managing a large number of POC trial candidates of comparable interest. It is found that, to be cost-effective, the best strategy is to conduct a number of small trials and set the bar high for continuing the development program. Section "Optimal Program and Portfolio-level Strategy" discusses the best program-level and portfolio strategy for managing a small group of POC trial candidates (for one or more drugs) under a fixed budget. Uncertainty about the relationship between a Phase II endpoint and a Phase III endpoint is incorporated into the discussion. Section "Optimal Strategy for a Single Seamless Phase II/III Trial" discusses optimal design of a single trial wherein a seamless Phase II/III design is used for illustration to further incorporate various other considerations. Section "Summary" concludes with a summary. Additional applications and extensions including development of personalized medicines and optimization of Phase III futility analyses can be found in Beckman et al. [2], Chen and Beckman [4], and Song and Chen [15].

Key Design Parameters

Consider a typical randomized Phase II POC trial of two arms with a 1:1 ratio (study drug versus placebo, or standard of care plus study drug versus standard of care plus placebo). Denote by δ the standardized effect size (treatment effect divided by standard deviation) of clinical interest with respect to a Phase II endpoint (an early efficacy endpoint or loosely a surrogate endpoint relative to clinical endpoint in Phase III). Denote by (α, β) the doublet of one-sided Type I error rate and Type II error rate (i.e., 1-power) of the trial. The total sample size for the trial is approximately

$$N = 4\left(Z_\alpha + Z_\beta\right)^2 / \delta^2 \tag{9.1}$$

where $Z_{(\cdot)}$ denotes the respective quantile of the standard normal distribution. When a time-to-event variable is the primary endpoint of interest, δ refers to the logarithm of hazard ratio, and N refers to the number of events. While the totality of data will be looked at closely, a Go decision to a Phase III confirmatory trial is generally made if the one-sided p-value from the POC trial is less than α favoring the study drug. Notice that the standard error for estimate of the treatment difference is $2/\sqrt{N}$ which is equal to the absolute value of $\delta/(Z_\alpha+Z_\beta)$ from the sample size formula, the cutoff point for the minimum empirical treatment difference (empirical bar) relative to δ in a Go decision (i.e., corresponding to one-sided p-value$<\alpha$) is $Z_\alpha/(Z_\alpha+Z_\beta)$. Clearly, the empirical bar increases when Type I error rate decreases or when Type II error rate increases.

In practice, δ is often chosen to make sure that the projected effect size in the Phase III endpoint, denoted by Δ, is clinically meaningful. We assume that $\breve{\varphi}$ is an unbiased estimate of the relative effect size between the two endpoints (Δ/δ) based on historical data. The estimate has an asymptotic normal density distribution with mean φ and variance υ^2 (i.e., $\phi\left(\breve{\varphi} \mid \varphi, \upsilon^2\right)$). In order for an early efficacy endpoint to be usable, the value of φ should be less than 1. Otherwise, it would take larger sample size or larger number of events to conduct a Phase II than a Phase III trial. This forfeits the purpose of using early efficacy endpoints in Phase II trials to make a Go–No Go (GNG) decision for drug development. Clearly, from formula (9.1), we can see that the relative sample size (a surrogate of relative cost) of the POC trial to the Phase III confirmatory trial is approximately proportional to φ^2, and the smaller the value of φ the more sensitive the early efficacy endpoint. The variance υ^2 implies the uncertainty of the predicted treatment effect in Phase III, and the larger the variance the greater the uncertainty. The value of υ^2 is not accounted for in section "Optimal General Strategy for Single POC Trials" for ease of development of a general portfolio-level strategy, but is accounted for in sections "Optimal Program and Portfolio-level Strategy" and "Optimal Strategy for a Single Seamless Phase II/III Trial". Estimation of φ and υ^2 is not the focus of this chapter. Interested readers in oncology may find good examples in Sargent et al. [13] and Tang et al. [20]. As a general guidance, they should be estimated from proper meta-analysis of multiple large randomized and controlled trials in similar disease settings. Each trial is considered as a unit in the meta-analysis. If data from only one large trial is available, each center in the trial may be considered as a unit and the bootstrap technique may be applied as appropriate for estimation. Interested readers may refer to Burzykowski et al. [3] and Whitehead [21] for more details.

Unless otherwise specified in section "Optimal Strategy for a Single Seamless Phase II/III Trial", we assume that the investigational arm has a prior probability p of being active with effect size δ and probability $1-p$ of being inactive. Prior information on drug activity, be it subjective or objective, is frequently cited by relevant decision makers in a drug development program. However, the information is rarely fully accounted for in the actual (mostly qualitative) decision process. It is not our focus to estimate p (or more generally the distribution of true effect size) although an estimate of 10–30 % for an investigational new drug seems reasonable (or possibly generous). Instead, our focus is on how to derive a quantifiable decision process under the same assumption as used by decision makers. We have found that the high level conclusions of this decision process are relatively insensitive to the estimation of p.

Optimal General Strategy for Single POC Trials

With the above setup, we will introduce the benefit–cost ratio in its basic form and derive a high-level strategy for portfolio management of a large number of POC trial candidates of comparable interest without clear distinction among each other. Without loss of generality, we assume that a Phase III confirmatory trial with 2.5 % one-sided type I error and 90 % power ($\beta=0.1$) will be initiated, seamlessly or sequentially, once a Go decision is made after the POC trial is completed. The relative sample size of the POC trial (denoted by SS2) to the Phase III confirmatory trial (denoted by SS3) is assumed to be λ (i.e., $\lambda=\mathrm{SS2}/\mathrm{SS3}$) when both are conducted at same type I/II error rates, where λ may slightly differ from φ^2. When the POC trial is conducted at (α, β), the sample size of the trial is then

$$\mathrm{SS2}(\alpha,\beta)=\lambda\times\mathrm{SS3}\times\left(Z_{\alpha}+Z_{\beta}\right)^2/\left(Z_{0.025}+Z_{0.1}\right)^2 \tag{9.2}$$

Table 9.1 Optimal design of Phase II POC trials under various scenarios when SS3 = 1,000

		Conventional design (α, β) = (0.025, 0.2)		Optimal design			
p (%)	λ (%)	SS2	1/BCR	(α, β)	SS2	Empirical Go bar (δ)	1/BCR
10	10	59	2,215	(0.02, 0.40)	51	0.88	2,144
10	20	118	3,149	(0.04, 0.44)	71	0.92	2,871
10	30	177	4,082	(0.05, 0.47)	83	0.95	3,442
30	10	59	1,384	(0.03, 0.41)	45	0.90	1,356
30	20	118	1,695	(0.05, 0.46)	61	0.94	1,573
30	30	177	2,007	(0.07, 0.49)	68	0.98	1,739
50	10	59	1,218	(0.03, 0.43)	38	0.91	1,193
50	20	118	1,405	(0.06, 0.48)	48	0.97	1,304
50	30	177	1,591	(0.08, 0.52)	52	1.03	1,385

Consider the following benefit–cost ratio with SS2(α, β) provided in Eq. (9.2).

$$\mathrm{BCR}(\alpha,\beta)=\frac{p(1-\beta)}{\mathrm{SS2}(\alpha,\beta)+\left[p(1-\beta)+(1-p)\alpha\right]\times \mathrm{SS3}} \tag{9.3}$$

The numerator of Eq. (9.3) is the probability of a true positive outcome from the trial. The two terms in front of SS3 represents the probability of a true positive outcome and of a false positive outcome, respectively. Therefore, the denominator overall represents the expected total sample size for the Phase II/III program. The benefit–cost ratio measures how much a patient contributes to development of an active drug, and its inverse measures how many patients it takes to develop an active drug. Because the final cost of a trial is generally tied to sample size, maximization of the benefit–cost ratio with respect to (α, β) is approximately equivalent to finding the optimal design parameters that maximize the return on investment. Notice that the development expense prior to Phase II is not included in the denominator for a rational decision (the fact that humans are prone to loss aversion is called *sunk cost fallacy* in behavior science).

Table 9.1 provides the optimal design of Phase II POC trials under various assumptions on p and λ covering a wide range of parameter space of practical interest. The sample size for Phase III (SS3) is fixed at 1,000 for illustration; the optimal design parameters are independent of the actual value of SS3. A conventional design for Phase II POC with (α, β) fixed at (0.025, 0.2) is provided for comparison. Under the conventional design, the empirical bar for a Go decision is 0.70δ; the sample size increases proportionally with λ. Under the optimal design, the type I error rate is low at around 5 % but the optimal type II error rate is high at around 45 %. The empirical bar for a Go decision is generally close to δ; the sample size increases when p decreases or when λ increases but is not as sensitive to p and λ as one might have expected. Not surprisingly, the expected number of patients it takes to develop an active drug (i.e., 1/BCR) is smaller under the optimal design than under the conventional design (especially so when λ is large). Reduced Phase II sample size is a main driver of improved efficiency. As a matter of fact, the analysis indicates that a Phase II POC trial should only cost 4–8 % that of a Phase III confirmatory trial to be cost-effective.

Some drug developers may have come to the same strategy as the above (i.e., more smaller POC trials with higher Go bars) purely based on intuition and experience. To our best knowledge, benefit–cost ratio analysis is the first quantitative analysis that provides theoretical support of this strategy. It is no secret to seasoned executive decision makers in large pharmaceutical companies that the empirical bars for Go decisions have to be high to reduce Phase III failure rate. The benefit–cost ratio analysis demonstrates, perhaps to statisticians' dismay, that the bar should be set as high as the target effect size, which on the other hand actually helps make statisticians' job a little easier because it is often confusing to team members without statistical background why 0.70δ is good enough for a Go decision when the real interest is in δ. (Actually, in order to raise the Go bar to δ under the conventional type I/II error rates of (0.05, 0.2) for a POC trial, the target effect size should be set at approximately 1.5δ.)

While a low type II error rate is always desirable, it results in fewer larger POC trials due to fiscal constraint and some worthy hypotheses will fall below the funding line as a consequence. The opportunity cost of missing POC trials that might have identified a true positive, due to running a small number of large POC trials under a fixed budget, has been termed *type III error*. While conventional statistics focuses on type I and type II errors only, a consideration of type III error is critical to identifying an optimal POC strategy [2]. Although some active indications will be missed due to the higher type II error, this is more than compensated for by the reduction in type III error inherent in having more shots on goal in the total program.

In some cases, a larger POC trial may be suggested by other considerations. A Phase II POC trial may have multiple primary endpoints, or address more than one hypothesis [2]. It may also have a safety objective that needs a certain number of patients to address, irrespective of the efficacy objective. Futility analyses will reduce the expected Phase III sample size, increasing λ and the corresponding optimal POC sample size. Despite all these and various other practical considerations, the general strategy of conducting more and smaller POC trials remains essential for drug development to be cost-effective.

Optimal Program and Portfolio-level Strategy

A drug development team often encounters the problem as how to appropriately allocate resources when there are more POC trials of interest than the budget can afford. These POC trials may be for the same drug (program level strategy) or for different drugs (portfolio level strategy). To best address this issue, we take additional considerations into account starting with uncertainty on relative effect size (φ) between a Phase III endpoint and a Phase II endpoint.

For a POC trial with treatment effect δ based on an early efficacy endpoint, the corresponding best estimate of treatment effect in the Phase III endpoint is $\breve{\varphi}\delta$ at the planning stage of the POC trial. Let α_3 be the one-sided type I error rate and β_3 be the one-sided type II error rate for the Phase III trial powered for detecting treatment effect Δ. Based on Eq. (9.1), the actual type II error rate for the Phase III trial is estimated to be

$$\breve{\beta}_3^* = f\left\{\breve{\varphi}\delta\left(Z_{\alpha_3} + Z_{\beta_3}\right)/\Delta - Z_{\alpha_3}\right\} \tag{9.4}$$

where $\Phi(\cdot)$ denotes the cumulative distribution function of a standard normal variable. The expected type II error rate for the Phase III trial is estimated to be

$$E\left\{\breve{\beta}_3^*\right\} = \int f\left\{\breve{\varphi}\delta\left(Z_{\alpha_3} + Z_{\beta_3}\right)/\Delta - Z_{\alpha_3}\right\}\phi\left(\breve{\varphi}\,|\,\varphi,\upsilon^2\right)\mathrm{d}\breve{\varphi} \tag{9.5}$$

where $\phi\left(\breve{\varphi}\,|\,\varphi,\upsilon^2\right)$ is the estimated density function of $\breve{\varphi}$.

Suppose there are k candidate POC trials. Let (α_{i2}, β_{i2}, δ_i, p_i, B_i, C_{2i}, C_{3i}, $\phi_i\left(\breve{\varphi}_i\,|\,\varphi_i,\upsilon_i^2\right)$) be the design parameters associated with the ith one ($i=1,\ldots,k$) whereas (α_{i2}, β_{i2}) are the type I/II error rates, δ_i is the target effect size, p_i is the estimated probability of drug being truly active, B_i is the benefit (commercial value or any other metric), C_{2i} is the cost for Phase II, C_{3i} is the corresponding cost for Phase III, and $\phi_i\left(\breve{\varphi}_i\,|\,\varphi_i,\upsilon_i^2\right)$ is the estimated density function of $\breve{\varphi}$. Consider the following general version of the benefit–cost ratio function (Eq. 9.3), wherein the numerator is adjusted with expected Phase III power ($1-E\left\{\breve{\beta}_3^*\right\}$):

$$\mathrm{BCR}\left(\alpha_{i2},\beta_{i2},i=1,\ldots,k\right) = \frac{\sum_{i=1}^{k} B_i p_i\left(1-\beta_{i2}\right)\left(1-E\left\{\breve{\beta}_{3i}^*\right\}\right)}{\sum_{i=1}^{k}\left\{C_{2i} + \left[p_i\left(1-\beta_{i2}\right) + \left(1-p_i\right)\alpha_{i2}\right]\times C_{3i}\right\}} \tag{9.6}$$

The goal is to find the optimal (α_{i2}, β_{i2}, $i=1,\ldots,k$) that maximizes (Eq. 9.6) subject to a constraint $\sum_{i=1}^{k} C_{2i} = C_2$ where C_2 is the total budget for POC.

We use the same oncology example as in Chen et al. [8] for illustration. There are three candidate POC trials in the example. For simplicity, we will ignore the benefit terms. We also assume that the cost for a POC trial or a Phase III trial is totally driven by the sample size, and further the cost for Phase III is the same (i.e., $C_{3i}=C_3$, i$=1,\ldots,3$). In terms of relative effect sizes for endpoints, $(\varphi_1,\upsilon_1^2)=(0.6,0.25)$, $(\varphi_2,\upsilon_2^2)=(0.6,0.5)$ and $(\varphi_3,\upsilon_3^2)=(0.5,1)$ so that the first two endpoints are equally sensitive but there is less uncertainty in first one, and the third endpoint is the most sensitive but also has the greatest uncertainty. The first two studies are believed to have higher chance of success ($p_1=p_2=0.3$) than the third one ($p_3=0.2$).

Each of the Phase III trials targets a 25 % hazard reduction (i.e., $\Delta=\log(1-0.25)$) in overall survival (OS), which translates into $\delta_i=-0.48$ ($i=1$, 2) for the endpoints

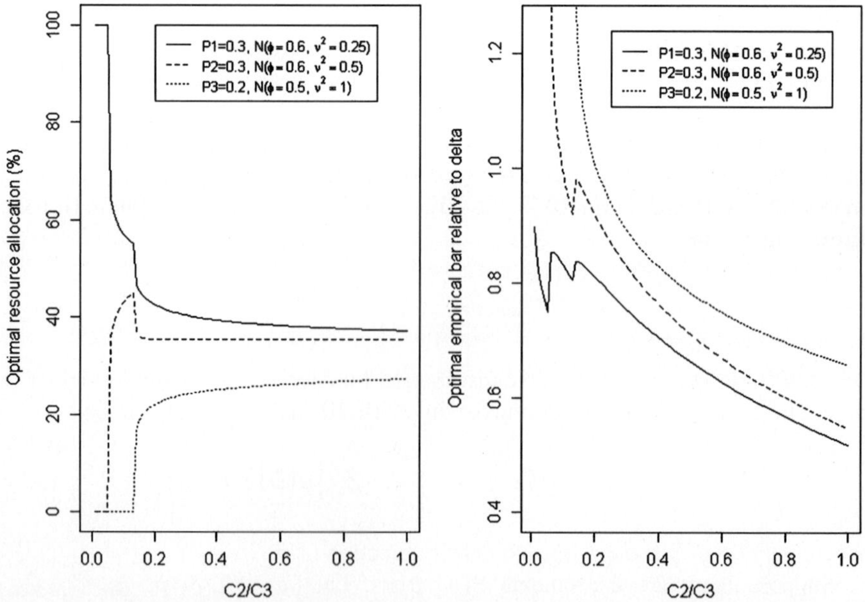

Fig. 9.1 Optimal resource allocation (*left* panel) and optimal empirical bars for a Go decision (*right* panel) for the three POC trials based on different endpoints

in first two trials (logarithm of hazard ratio in progression-free-survival or PFS) and $\delta_3 = -0.58$ the endpoint in the third trial (standardized tumor size change at Cycle 2). In comparison to a phase III trial with type I/II error rates (0.025, 0.1) which typically require 800–1,000 patients, a POC trial with same type I/II error rates would require about 238 patients to target $\delta = -0.48$ using PFS and 126 patients to target $\delta = -0.58$ using standardized tumor size change at Cycle 2.

With this setup, we try to find the optimal (α_{i2}, β_{i2}, $i = 1, \ldots, 3$) that maximizes (Eq. 9.6) subject to the constraint $\sum_{i=1}^{3} C_{2i} = 220$, the overall budget for POC. Once optimal (α_{i2}, β_{i2}, $i = 1, \ldots, 3$) are obtained for each trial, optimal sample size and empirical bar for a Go decision follow. Notice that, because the total sample size for the three POC trials is constrained at 220, the effective number of unknown parameters is reduced from 6 to 5 (R code available upon request).

Figure 9.1 shows optimal resource allocation (left panel) in terms of percentage of total sample size (out of 220 patients) and optimal empirical bars for a Go decision to Phase III (right panel) as a function of the ratio of overall Phase II budget ($C_2 = 220$ patients) to the sample size for an individual Phase III (C_3). The POC trial that uses PFS as the primary endpoint and has the greatest certainty in relative effect size always has the highest priority. Indeed, this is the only trial that should be conducted if the C_2/C_3 ratio is less than 0.06 (i.e., if a Phase III trial takes approximately 3,500 patients). When the ratio is between 0.06 and 0.13, the second POC trial based on PFS should also be considered. When the ratio is greater than 0.13, all three POC trials should be conducted. In particular, if the sample size for one Phase III trial is around 800 (typical sample size in first line lung cancer for targeting a 25 % hazard reduction in OS) or C_2/C_3 is around 0.275, approximately 40, 35 and 25 % of the resources should be allocated to the three respective trials. In this case, instead of somewhat arbitrarily choosing two from the three POC candidates and conducting each at greater sample size, the benefit–cost ratio analysis suggests a more opportunistic approach by conducting all three at smaller sample size. The two POC trials based on PFS would have approximately 88 and 77 patients (62 and 54 PFS events), and the one based on tumor size change would have approximately 55 patients. The respective empirical bars are 0.77, 0.85, and 0.91 relative to target effect size. As might be expected, in general, the empirical bar generally increases with sample size for Phase III (or decreases with C_2/C_3), and the larger the POC trial (or more upfront commitment) the lower the empirical bar. These results are sensible and consistent with intuition. However, intuition alone will not be able to pinpoint the optimal decision points.

Optimal Strategy for a Single Seamless Phase II/III Trial

Seamless Phase II/III designs are often considered to accelerate a development program. The GNG decision for the transition has to be made with limited Phase II data. It is a fine line to walk between benefit and cost. We use a real example from the development of a drug candidate for platinum resistant ovarian cancer patients to illustrate how benefit–cost ratio analysis helps with the decision making process [19].

Description of Study Design

The primary hypothesis of the Phase II/III confirmatory trial is as follows: the test drug is non-inferior to the comparator (a standard of care chemotherapy) in terms of OS at the 1.1 hazard ratio margin and superior to the comparator in terms of safety profile, OR the test drug is superior to the comparator in terms of OS.

The trial is operationally seamless. Enrollment is potentially seamless between Phase II and Phase III but only Phase III data is used in the final analysis. The design of the motivating example is shown schematically in Fig. 9.2 with the GNG bars derived in next sections. In the Phase II portion, patients will be randomized to three treatment groups with equal allocation: test drug at high dose, test drug at low dose, and control. The primary endpoint for Phase II is PFS. Phase II enrolls about 210 patients and completes after 135 PFS events have been observed to have sufficient power for each dose of the test drug to demonstrate superiority to the control in terms of PFS. The primary endpoint of Phase III is OS. Phase III enrolls about 720 patients and completes after 508 deaths have been observed to have sufficient power to demonstrate that the test drug is non-inferior to the control drug. This sample size also provides sufficient power (>95 %) to demonstrate that the test drug is superior to the control in terms of the event rate for a safety endpoint.

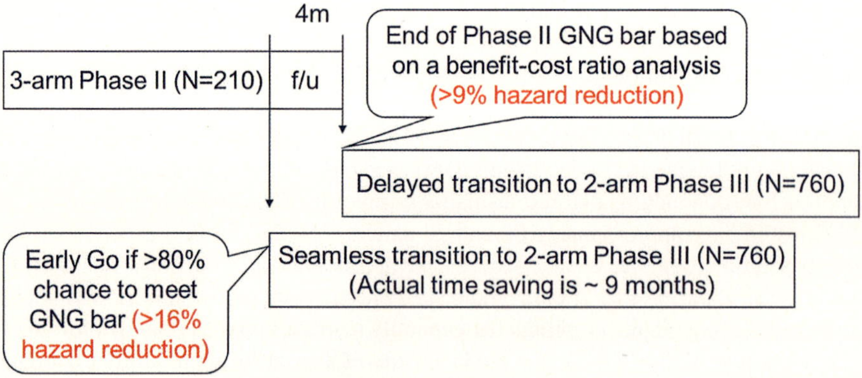

Fig. 9.2 Flowchart of the seamless Phase II/III study in platinum resistant ovarian cancer patients

In order to realize seamless transition, an interim analysis will be conducted in Phase II. The enrollment of Phase II will close when it is predicted that approximately 4 months after this time point there will be 135 PFS events. The interim analysis will take place approximately 1 month before the accrual completion. The purpose of this interim analysis is to determine whether Phase III enrollment can be initiated before final data of Phase II is available. If a Go decision is made, one arm of the test drug along with the control arm will be carried to Phase III. If a Go decision cannot be made at the interim analysis, Phase III will be on hold and a final decision will be made at end of Phase II. The Go criterion at this interim analysis is to have at least 80 % conditional power to meet the GNG bar at the final analysis of Phase II. Since it will take about 1 month to conduct the interim analysis and make a decision, the timing of this interim analysis is chosen so that Phase III accrual will potentially start seamlessly when Phase II accrual completes.

Weighted Estimate of OS Effect in Phase II

A common practice in drug development is to make GNG decision only based on a Phase II endpoint. However this approach often causes heated debate as what role the clinical endpoint data plays. In oncology, this often leads to a vague conditional requirement of a "positive OS trend" before a Go decision can be made. In Chen and Sun [7], it is proposed to combine the PFS data and OS data for decision making so that no information is wasted and a decision rule can be prespecified without ambiguity. The same approach can be used to combine data across multiple endpoints, not only in oncology but also in any other therapeutic areas.

In our motivating example, the relative effective size (φ) between OS and PFS (in log-hazard-ratio scale) is estimated to be 0.6. It implies that the treatment effect in

OS is 60 % of that in PFS, which represents a reasonable estimate based on published data of a variety of solid tumors in recent years. The relative effect size suggests a hazard ratio of 0.69 in PFS correspond to a hazard ratio of 0.8 in OS. To adequately account for the uncertainty in relative effect size, we assume that it has a standard deviation of 0.2. This assumption covers a wide range of effect size ratios seen in the literature. With this variability, a hazard ratio of 0.69 in PFS translates into an OS effect with 95 % confidence interval (0.69, 0.93).

A weighted method is used to estimate the OS effect. It combines the predicted OS effect from the observed PFS effect ($\breve{\varphi}\breve{\delta}_{\mathrm{PFS}}$) and the observed OS effect OS ($\hat{\Delta}_{\mathrm{OS}}$), both in log-hazard-ratio scale (see below), as follows:

$$S = -\left(w\hat{\Delta}_{\mathrm{OS}} + (1-w)\breve{\varphi}\breve{\delta}_{\mathrm{PFS}} \right) \tag{9.7}$$

With minus sign on the right-hand side, S is an approximate measure of hazard reduction, a parameter clinical researchers are more familiar with. Since the number of OS events in Phase II is relatively small compared to the number of PFS events a weight of 0.15 (i.e., $w=0.15$) is given to the observed OS effect in Phase II, and a weight of 0.85 is given to the predicted OS effect. This weight approximately minimizes the variance of S when the true treatment effect is in the parameter space of interest while the actual numbers of PFS and OS events are in the expected range. The correlation between $\breve{\delta}_{\mathrm{PFS}}$ and $\hat{\Delta}_{\mathrm{OS}}$, and the variance of $\breve{\varphi}$ are all incorporated into the variance estimate of S. (See Chen and Sun [7] and Sun and Chen [19] for technical details on characteristics of the test statistics.)

In the next section, we discuss what value of S will constitute a GNG criterion between Phase II and Phase III using the same technique as developed in the previous sections.

A Benefit–Cost Effective GNG Criteria from Phase II to Phase III

The test drug is assumed to have equal probability of being superior to control (hazard ratio = 0.8), equivalent to control (hazard ratio = 1) and inferior to control (hazard ratio = 1.1). Let the Go criterion from Phase II to Phase III at end of Phase II be $S > C$, where S is defined in Eq. (9.7). The probability of making a Go decision under each hazard ratio assumption easily follows. In this example, the Phase III is successful in two scenarios: (1) Superiority in efficacy is demonstrated; (2) Only non-inferiority in efficacy and superiority in safety are demonstrated. The chance of regulatory approval and commercial value are different in these two scenarios. In our example, stakeholders and experts believe the relative chance of approval from health authorities is 2:1 for scenario 1 vs. scenario 2, and the corresponding relative commercial value is 5:1. Therefore the relative regulatory risk adjusted benefit of the two scenarios is 10. Further, it is estimated that at the time the GNG

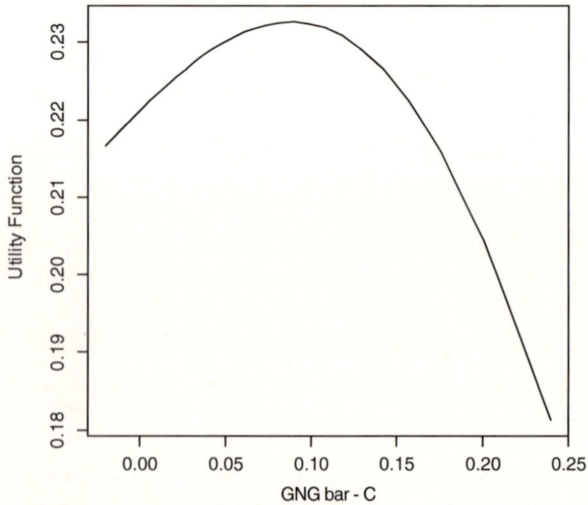

Fig. 9.3 Illustration of how the utility function (benefit–cost ratio) changes with the GNG criterion (C) in the Phase II/III study in platinum resistant ovarian cancer patients

decision is made for seamless transition the cost of Phase II plus Phase III preparation is approximately 40 % that of the remaining cost for Phase III.

Now, consider the following benefit–cost ratio function, whereas $\Pr(S>C|\text{Super})$, $\Pr(S>C|\text{Equiv})$, and $\Pr(S>C|\text{Inf})$ are the probability of Go under the superiority, equivalence, and inferiority assumptions respectively, $Q(\text{Super})$ and $Q(\text{Equiv})$ are the power and regulatory risk adjusted benefits under the superiority and equivalence assumptions. The adjusted benefit under a hazard ratio assumption is the sum of the power of superiority multiplied by 10 and the power of non-inferiority. An inferior drug does not have any benefit. Notice that sum of the three terms in bracket of the denominator is the probability of going to Phase III, and the denominator represents the multiples of the remaining cost for Phase III.

$$\text{BCR}(C)=\frac{\frac{1}{3}\Pr(S>C\,|\,\text{Super})Q(\text{Super})+\frac{1}{3}\Pr(S>C\,|\,\text{Equiv})Q(\text{Equiv})}{0.4+\left[\frac{1}{3}\Pr(S>C\,|\,\text{Super})+\frac{1}{3}\Pr(S>C\,|\,\text{Equiv})+\frac{1}{3}\Pr(S>C\,|\,\text{Inf})\right]} \quad (9.8)$$

The optimal GNG bar C is obtained by maximizing the above benefit–cost ratio, and it turns out to be $C=0.09$. Roughly speaking, this corresponds to a 9 % hazard reduction based on the joint estimate of the OS effect by using both PFS data and OS data from Phase II as well as the estimate of relative effect size based on historical data. Figure 9.3 illustrates how the benefit–cost ratio changes with C, which decreases when it moves farther away from the optimal value. This is typical in a benefit–cost ratio analysis; the optimal design is unique and optimality is mathematically global. With the optimal GNG bar for the end of Phase II data obtained, we can back-calculate the GNG bar at the interim analysis of Phase II for a seamless transition to Phase III. It turns out that the bar is approximately a 16 % hazard reduction (i.e., 80 % chance that the hazard reduction is at least 9 % at the end of Phase II if its 16 % at the interim analysis). To mitigate the risk of using a wrong assumption on relative effect size, $\breve{\Delta}_{OS}$ and $\breve{\varphi}\breve{\delta}_{PFS}$ are compared for consistency before a final decision is made [19].

Summary

In this chapter, we have defined a benefit–cost ratio function for measuring efficiency of Phase II POC trials. The idea is applied to derive optimal trial-level, program-level, and portfolio-level design strategies so that readers with different statistical background and perspective ranging from project statisticians, development team leaders, to executive decision makers can appreciate the underlying concepts and immediately apply them to contemporary late-stage drug development.

In general, conducting more and smaller POC trials and setting the GNG bars higher is more cost effective than conducting a few larger POC trials with lower GNG bars. When determining resource allocation in portfolio management, trials with lower perceived probability of success or an imperfect early endpoint may still be considered unless Phase III cost is prohibitively high. However, less resource should be allocated to these trials and the GNG bars should be set higher. The paradigm allows one to calculate an optimal resource allocation for POC trials of varying value, including determining whether a trial should not be funded.

Seamless Phase II/III designs have great promise in accelerating a development program. A detailed example is presented utilizing the benefit–cost ratio analysis and a weighted estimate of treatment effect based on several endpoints to specify an objective, optimal GNG decision rule.

Objective specification of study designs and GNG decision rules determined using a benefit–cost ratio analysis is the key to maximizing return on investment in Phase II POC trials.

References

1. Anderson KA (2006) Optimal spending functions for asymmetric group sequential designs. Biom J 48:1–9
2. Beckman RA, Clark J, Chen C (2011) Integrating predictive biomarkers and classifiers into oncology clinical development programs: an adaptive, evidence-based approach. Nat Rev Drug Discov 10:735–749
3. Burzykowski T, Molenberghs G, Buyse M (2005) The evaluation of surrogate endpoints. Springer, New York
4. Chen C, Beckman RA (2009) Optimal cost-effective designs of proof of concept trials and associated go-no go decisions. J Biopharm Stat 19(2):424–436
5. Chen C, Beckman RA (2009) Hypothesis testing in a confirmatory phase III trial with a possible subset effect. Stat Biopharm Res 1(4):431–440

6. Chen C, Beckman RA (2009) Optimal cost-effective go-no go decisions in late-stage oncology drug development. Stat Biopharm Res 1(2):159–169
7. Chen C, Sun L (2011) Quantification of PFS effect for accelerated approval of oncology drugs. Stat Biopharm Res 3:434–444
8. Chen C, Sun L, Chih C (2013) Evaluation of early efficacy endpoints for proof-of-concept trials. J Biopharm Stat 23:413
9. Halpern J, Brown BW, Hornberger J (2001) The sample size for a clinical trial: a Bayesian-decision theoretic approach. Stat Med 20:841–858
10. Leung D, Wang Y (2001) A Bayesian decision approach for sample size determination in phase II trials. Biometrics 57:309–312
11. O'Hagan A, Stevens JW (2001) Bayesian assessment of sample size for clinical trials of cost-effectiveness. Med Decis Making 21:219–230
12. Patel NR, Ankolekar S (2007) A Bayesian approach for incorporating economic factors in sample size design for clinical trials of individual drugs and portfolios of drugs. Stat Med 26:4976–4988
13. Sargent D, Wieand S, Haller DG et al (2005) Disease-free survival (DFS) vs. overall survival (OS) as a primary endpoint for adjuvant colon cancer studies: individual patient data from 20, 898 patents on 18 randomized trials. J Clin Oncol 23:8664–8670
14. Simon R (1989) Optimal two-stage designs for phase II clinical trials. Control Clin Trials 10:1–10.7
15. Song Y, Chen C (2012) Optimal strategies for developing a late-stage clinical program with a possible subset effect. Stat Biopharm Res 4(3):240–251
16. Stallard N (1998) Sample size determination for phase II clinical trials based on Bayesian decision theory. Biometrics 54:279–294
17. Stallard N (2003) Decision-theoretic designs for phase II clinical trials allowing for competing studies. Biometrics 59:402–409
18. Stallard N, Whiehead J, Cleall S (2005) Decision-making in a phase II clinical trial: a new approach combining bayesian and frequentist concepts. Pharm Stat 4:119–128
19. Sun L, Chen C (2012) Advanced application of using progression-free survival to make optimal go-no go decision in oncology drug development. In: JSM proceedings biopharmaceutical section, San Diego, August 2012
20. Tang PA, Bentzen SM, Chen EX, Siu LL (2007) Surrogate end points for median overall survival in metastatic colorectal cancer: literature-based analysis from 39 randomized controlled trials of first-line chemotherapy. J Clin Oncol 25:4562–4568
21. Whitehead A (2002) Meta-analysis of controlled clinical trials. Wiley, Chichester

Chapter 10
Portfolio Optimization of Therapies and Their Predictive Biomarkers

Robert A. Beckman and Cong Chen

Introduction: Predictive Biomarkers, Value and Risks

Predictive biomarkers, or "response identification biomarkers," are molecular or other characteristics of patients or of their tumors which may predict whether or not they will benefit from therapy. Predictive classifiers may be constructed from one or more predictive biomarkers. With increasing understanding of cancer and its therapies, the number and importance of predictive biomarkers and classifiers is increasing.

Cancer therapy development currently carries significant risk, a major portion of which is due to heterogeneity between patient's tumors. As a result of molecular heterogeneity between patients with similar tumor histology, a given therapy may benefit only a minority of the patients. The signal from this minority of patients may be diluted by noise from the other patients in a trial in an unselected population, resulting in a negative study. For example, trastuzumab was studied in metastatic breast cancer patients whose tumors overexpressed its target, the Her2-neu protein [16, 35]. This therapy has been very successful in benefiting approximately 20 % of

R.A. Beckman, M.D. (✉)
Center for Evolution and Cancer, Helen Diller Family Cancer Center, University of California at San Francisco, San Francisco, CA, USA

Oncology Clinical Research, Daiichi Sankyo Pharmaceutical Development, Edison, NJ, USA

1045 Hereford Drive, Blue Bell, PA 19422, USA
e-mail: eniac1915@gmail.com

C. Chen, Ph.D.
Biostatistics and Decision Research Sciences, Merck Research Laboratories, Upper Gwynedd, PA, USA
e-mail: cong_chen@merck.com

Z. Antonijevic (ed.), *Optimization of Pharmaceutical R&D Programs and Portfolios: Design and Investment Strategy*, DOI 10.1007/978-3-319-09075-7_10

metastatic breast cancer patients, but if the clinical trial had been done in an unselected metastatic breast cancer population it almost certainly would have been negative.

The small average benefit seen in cancer clinical trials may be due to this dilution of the signal that is really present only in a subset of the patients. Detecting such small clinical effect sizes requires large, expensive clinical trials. The high risk and high cost of therapy development raises the cost of therapies. The high cost and small average benefit leads to very low relative value of many oncology medicines, when judged by the cost of a Quality Adjusted Life Year (QALY). National health authorities and payors are increasingly demanding value for medicines. For example, the UK National Institute for Clinical Excellence (NICE) requires therapies to provide a QALY for less than or equal to 30,000 British pounds, a threshold achieved by few oncology therapeutics.

Predictive biomarkers have the potential to increase the benefit to patients by matching each patient subset to its optimal therapy, and to reduce the risk and the cost of cancer therapy development. Examples of successful applications of predictive biomarkers include her2neu expression for trastuzumab therapy of breast cancer [16, 35], sensitizing mutations in the epidermal growth factor receptor (EGFR) gene for gefitinib and erlotinib therapy of non-small-cell lung cancer [26, 27], ras wild type status for therapy of colorectal cancer with anti-EGFR therapy using cetuximab or panitumumab [1, 8, 24, 37], alk translocations for crizotinib therapy of lung cancer [32], and V600E mutations for vemurafenib therapy of melanoma [10]. There can be little doubt that predictive biomarkers are essential for further progress in oncology.

However, predictive biomarkers do not always work, and when they do not, they simply add cost, complexity, and time to drug development. A particularly instructive example is the discovery that the anti-EGFR therapy cetuximab benefits patients who do not express the EGFR receptor as judged by immunohistochemistry. This may have been due to insufficient assay sensitivity, loss of antigen upon storage, chance sampling in a negative region of a heterogeneous tumor, or evolution of the tumor between the time when the sample was taken and the time when the patient was treated [29, 38]. These pitfalls and others are common to all predictive biomarkers in real clinical situations. Predictive biomarker performance must be robust to these issues, and this can only be tested clinically.

In general, predictive biomarkers are developed from preclinical models which have been designed to emphasize certain features of a biological mechanism, whereas in the clinical situation greater biological complexity is in play. The ability of a preclinical result to translate to the clinic is often unknown.

The drug development team may include members who developed the predictive biomarker hypotheses, and they may have difficulty objectively assessing the clinical applicability of their findings, leading to overemphasis on the hypotheses in subsequent development. Publication bias, resulting in more prominent publication of biomarker successes than failures, further clouds objective assessment of the value of predictive biomarkers.

The use of predictive biomarkers involves various costs and challenges. Significant resources must be invested in order to discover biomarkers, develop assays for biomarkers, and formally develop a "companion diagnostic assay" which meets regulatory requirements for co-approval with the therapy as a means of selecting patients [23]. Patient selection requires the availability of suitable diagnostic tissue, and patients who have insufficient diagnostic tissue will be ineligible for the trial, leading to recruitment difficulties.

The enormous potential and significant costs and risks of predictive biomarker-driven development have led to a range of attitudes ranging from high enthusiasm to significant skepticism [17, 30]. The latter publication expresses the point of view of skeptics well: "Whereas 'wins' have occurred here, … most attempts to identify such biomarkers have been nothing more than expensive fishing expeditions. Drug response is multifactorial; patient populations are heterogeneous; potential markers are innumerable; and scientific underpinnings to marker development are imperfect."

These legitimate issues and concerns may threaten the progress in a field whose value and promise is increasing daily with increased knowledge of molecular oncology. Beckman, Clark, and Chen [5] presented an adaptive, data-driven approach to integration of predictive biomarkers in oncology development, based on several years of discussion with experts from discovery, preclinical, clinical, statistical, regulatory, and commercial functions at several pharmaceutical and biotechnology companies. The approach steers a middle course between uncritical enthusiasm and harsh skepticism. This chapter reviews this approach within the context of a single indication, including original clinical trial designs.

However, when considering the portfolio view we must consider not only one predictive biomarker and its integration within the development of a single indication. Rather the oncology portfolio should be viewed as a collection of assets in two primary classes: potential therapies and potential predictive biomarkers, both of which are characterized by uncertainty. Given a finite budget, the value and risks of each of these assets must be considered in the portfolio optimization problem, as the budget will generally be insufficient for full investment in all. The chapter concludes with a consideration of this portfolio level problem, including relevant interdependencies between therapeutic and predictive asset classes.

The chapter is designed for readers with all levels of mathematical background; mathematical details can be found in the primary references.

Core Principles of Predictive Biomarker Evaluation

Two core principles underlie this chapter. The first is clinical validation of predictive biomarkers. This first principle is essential for determining the value of predictive biomarkers. The second central principle is optimization of development efficiency, where efficiency is defined as throughput of successful product/indications launches per late development patient utilized or dollar spent. This second principle leads directly to maximizing the value of a portfolio of putative therapies and putative predictive biomarkers.

Core Principle 1: Clinical Validation of Predictive Biomarkers

The principle of clinical validation has two corollaries: (1) Biomarker negative patients must be included in early exploratory efficacy testing, and in some cases in later confirmatory pivotal testing. (2) A single primary predictive biomarker hypothesis must be chosen for formal statistical testing against clinical benefit in Phase 2.

Testing of biomarker negative patients presents an ethical dilemma, if we believe that the therapy may not benefit them. The ethical standard for including them is *equipoise*, defined as a state of uncertainty regarding which arm of the randomized clinical trial is optimal for the biomarker negative patients. As discussed above, equipoise tends to be underestimated on drug development teams for a number of reasons including the presence of people on those teams responsible for preclinical development of the biomarker hypothesis; the fact that the hypothesis has probably been illustrated in a homogeneous, controlled experimental setting specifically designed for such an illustration; and publication bias leading to overestimation of probability of success of predictive biomarker hypotheses. In the setting of equipoise, there may be an ethical issue with denying a biomarker negative patient the opportunity to benefit from an experimental therapy when he/she has few other options. Anti-EGFR therapies illustrate the issues. Erlotinib and gefitinib offer benefit to unselected populations, although it is less than for those who have sensitizing mutations of EGFR [26, 27, 33]. As discussed above, cetuximab even provides a degree of benefit to patients who are "EGFR negative" by immunohistochemistry. Fridlyand et al [21] list four factors which may increase or decrease equipoise with respect to a predictive biomarker:

1. A clear cutoff defining mutually exclusive groups as is seen in association with mutations or gene amplifications leads to greater certainty than an uncertain cutoff, as would be true for a continuous biomarker (gene expression) or a categorical biomarker (1+,2+,3,4+ as in immunohistochemistry).
2. A definite, single, and specific mechanism of action leads to more certainty than an unknown or a pleiotropic mechanism of action.
3. Preclinical data indicating that the predictive biomarker has high specificity and sensitivity, as opposed to just one or the other, leads to more certainty.
4. Certainty can be enhanced by a "class effect," i.e., prior knowledge of therapies with a similar mechanism of action.

Nonetheless we do not generally favor enrolling biomarker negative patients onto single agent trials of targeted therapy versus placebo. Rather, we favor randomized add-on designs in which all patients receive a backbone therapy. Thus the control arm is backbone therapy plus placebo and the experimental arm is backbone therapy plus experimental therapy. Ordinarily, backbone therapy is the standard of care (SOC) for that condition, but in cases where there is no SOC another therapy may be substituted as backbone. In some cases, the SOC therapy may be antagonistic to the experimental agent. This should be carefully ruled out to the extent feasible by preclinical in vitro and in vivo experiments. If SOC is antagonistic to the experimental agent, another indication should be chosen to test the biomarker hypothesis in which combination with SOC is feasible, that is in which another background therapy that is not antagonistic to the experimental therapy can be logically selected.

The second corollary of the need for clinical validation is prospective selection of a single primary predictive biomarker hypothesis for formal statistical testing in the randomized phase 2 study. As discussed in section "Efficiency Optimized Biomarker Stratified Randomized Phase 2 Study," the single primary predictive biomarker hypothesis is tested with equal priority as the drug itself. In too many cases, we see either a large number of biomarker subsets prospectively chosen for investigation in Phase 2, or still worse a large number of biomarkers are applied retrospectively after the clinical data are available, looking for a "win." And it is highly likely that a "win" can be found if a very large number of biomarker subsets are investigated (especially retrospectively), but such a win is usually a false positive. The chance of a false positive goes up in nearly direct proportion to the number of subsets investigated, a phenomenon which is termed the *multiple comparisons problem.* Such false positives result in negative phase 3 studies at great expense, as well as disillusionment with predictive biomarkers and personalized medicine.

The primary predictive biomarker hypothesis is selected based on in vitro preclinical data in cell lines; fundamental biochemical studies of mechanism of action; preclinical in vivo studies including xenograft models and syngeneic models including genetically engineered mouse models; molecular epidemiology research on biomarker prevalence in human tumor tissues; efficacy and pharmacodynamic data from Phase 1 studies in patients and in healthy normal volunteers, and Phase 2a signal seeking studies that look for tumor shrinkage and other signs in unrandomized single agent and combination cohorts. The prevalence of the biomarker in human tumor tissues is a critical piece of information. If the biomarker negative group is very small, eliminating these patients from the treated group may not be cost effective when the cost and error rate of the predictive test are considered. If the biomarker positive group is very small, it may be difficult to enroll the required clinical trials unless the benefit is dramatic, and pharmaceutical and biotechnology firms will also have to decide if the small subgroup is worth their investment.

The process of choosing the primary predictive biomarker hypothesis is far from foolproof. There may be several predictive biomarker hypotheses supported by nearly equivalent levels of evidence. These alternative biomarker hypotheses may also be explored in Phase 2, but as exploratory endpoints. If the primary predictive biomarker hypothesis proves to be false, but one of the exploratory alternative predictive hypotheses proves to be true, this is a lower level of evidence. It will be necessary to repeat Phase 2 and prospectively test the former alternative hypothesis as a new primary predictive biomarker hypothesis.

From a portfolio perspective, the primary predictive biomarker hypothesis is analogous to the lead compound, and the alternative biomarker hypotheses are analogous to backups. The uncertainty in choosing the primary predictive biomarker hypothesis is mirrored by the uncertainty in choosing the lead.

The need to occasionally iterate more than once through Phase 2 should not be seen as failure if what was learned the first time through enables a higher probability of success in subsequent iterations. We term such iterations *productive iterations* in recognition of the fact that they represent progress, and are a necessary consequence of the complexity of oncology.

Core Principle 2: Optimizing Development Efficiency

The concept of optimizing development efficiency is treated in detail in the chapter by [39], but as it is central to the material in this chapter, we also discuss it briefly here.

Chen and Beckman [11–13] investigated the optimal sizing of Phase 2 proof of concept (POC) trials and the optimal Go–No Go criteria for proceeding from Phase 2 to Phase 3 as judged by development efficiency, which they defined using a benefit–cost ratio (BCR) for late development (Phase 2 and 3 randomized trials). Benefit was defined as the risk adjusted number of therapies *correctly* identified for Phase 3 development (i.e., true positives only, not false positives). Cost was defined as the number of risk adjusted patients utilized in Phase 2 and Phase 3 studies. "Risk adjustment" means that the results were adjusted for the false positive and false negative rates of the Phase 2 POC study (Type I and Type II error rates, respectively) and the false negative rate of the Phase 3 study. False negative POC results lead to a missed opportunity: a therapy that would have worked is rejected for Phase 3 development. False positive POC results lead to an unwise expenditure: patients and financial resources are spent for Phase 3 on a therapy that is destined to fail. False negative Phase 3 results lead to missed opportunity even after correct identification of a true positive therapy by the POC trial (Fig. 10.1).

Fig. 10.1 (continued) cost of Phase 3. The chance of reaching any of the *rectangles* is the product of all the *arrows* on the path to that *rectangle*. The risk adjusted benefit is the sum of the benefits in each *rectangle*, each multiplied by the chance of reaching that *rectangle*. The risk adjusted cost is the sum of the costs (in patients) in each *rectangle*, each multiplied by the chance of reaching that *rectangle*. The efficiency is the risk adjusted benefit divided by the risk adjusted cost. (**a**) A single traditional POC trial with 80 % power for the minimum clinically significant effect size δ is performed, at a cost of 160 patients. The risk adjusted benefit is 0.24 true positives developed through phase 3. The risk adjusted cost is 346 patients. The efficiency is 6.94×10^{-4}. (**b**) Two small POC trials (in different indications of equal merit), each with approximately 60 % power for the minimum effect size **δ**, are performed at a cost of 80 patients each (only one is shown), using the same fixed POC budget as in **a**. The risk adjusted benefit is 0.18 true positives developed through phase 3 for each POC trial, for a total of 0.36 true positives developed through phase 3. The risk adjusted cost, including both small POC trials and all the Phase 3 trials resulting from them, is 460 patients. The efficiency is 7.83×10^{-4}, 13 % higher than that of the program with a single traditional POC trial (**a**). The second indication would not have been explored in **a** due to the fixed POC budget. This type III error reduces the efficiency of **a**, which is less efficient despite its lower type II error per POC trial

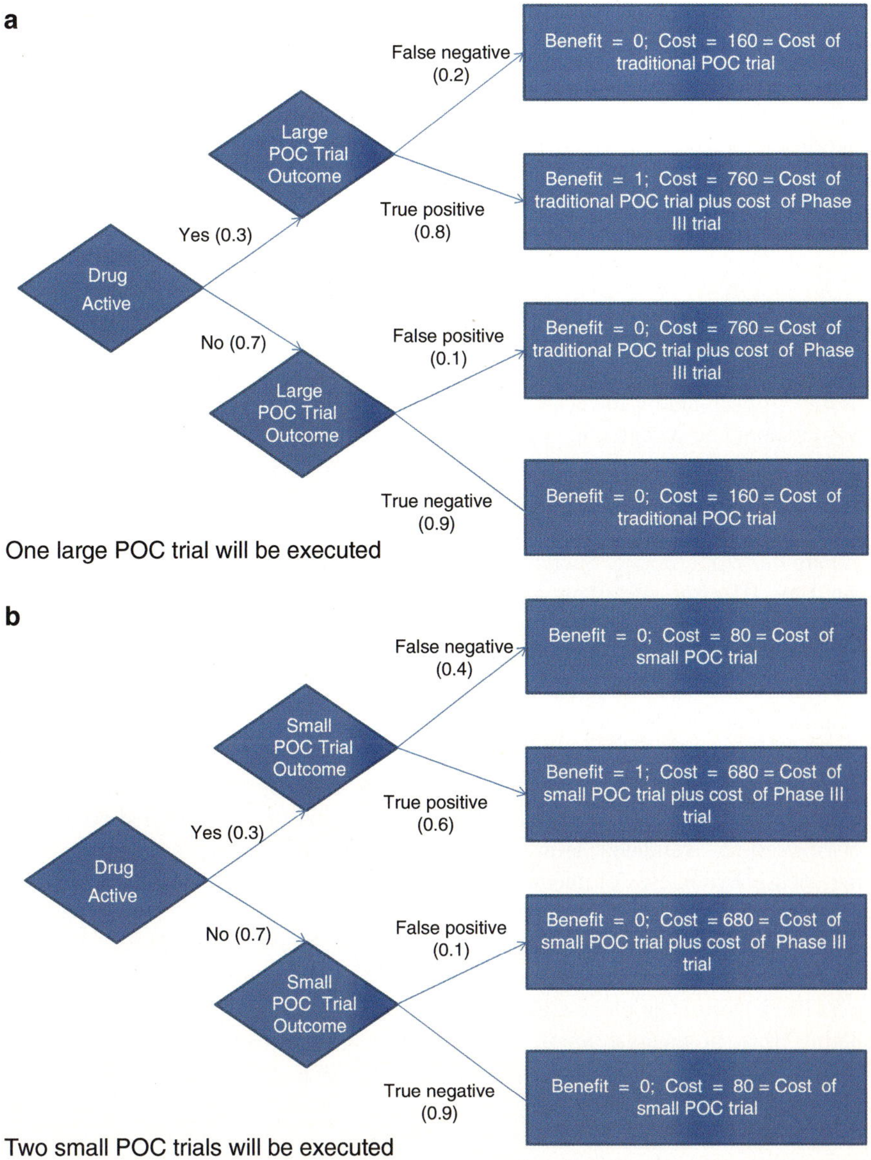

Fig. 10.1 Diagrams show the logic of a POC program for a drug with a probability of being active of 30 % and a minimally clinically significant PFS hazard ratio of 0.6. We compare one large POC trial to two small ones. All possible development outcomes are shown. Each *arrow* is labeled with its associated outcome and probability of occurrence. We assume the drug requires a 600 patient phase 3 study. On the left is a decision *diamond* reflecting the unknown effectiveness of the drug. In the upper branch, the drug is truly active; in the lower branch, it is truly inactive. An active drug proceeds to the upper middle *diamond*. The POC trial either produces a false negative (probability equal to the type II error rate β) or a true positive (probability equal to the study power $(1-\beta)$). An inactive drug proceeds to the lower middle *diamond*. The POC trial either produces a false positive (probability equal to the type I error rate α) or a true negative (probability equal to $(1-\alpha)$). The benefits and costs of the associated outcomes are shown in the *rectangles* on the right. Only true positives produce a benefit. Both true positives and false positives are associated with the additional

The studies of POC trial optimization assumed a fixed, finite POC budget, and a large portfolio of POC trial opportunities that greatly exceeds the available budget. This is in agreement with the authors' experience at several pharmaceutical and biotechnology firms, and is not surprising given the large number of oncology indications and possible standard and experimental drug combinations. It is assumed that if a POC trial is positive, resources for the subsequent Phase 3 trial will be made available, also generally in agreement with experience.

Defining optimal POC trial size and Go–No Go criteria for various therapy trials in the face of a limiting budget is an important portfolio optimization problem. Smaller POC trials have higher Type I and Type II error rates. However, smaller POC trials also allow more therapeutic hypotheses to be tested within the fixed POC budget.

Surprisingly, it was found that the optimal POC trial is smaller than what is traditionally used, and a lower power (higher Type II error rate) than the traditional power should be accepted in order to allow more hypotheses to be tested. When the higher traditional power is chosen, the Type II error rate is lower, but there is an opportunity cost in that fewer POC studies can be funded within the fixed budget. There is a chance that a POC study that would have led to a true positive would be among those not funded. Such an event was termed a "Type III error" [5]. Notably, the Type III error rate plays a significant role in determining the optimal POC design. In a typical example, it was shown that two POC studies at 60 % power, were 10–30 % more efficient than a single POC study at 80 % power, depending on the conditions (Fig. 10.1). The optimal power, lower than that traditionally recommended, is termed "Chen–Beckman power."

The above results assume that all therapeutic hypotheses are of equal merit, with equal probabilities of success (POS) prior to Phase 2 as judged by preclinical and Phase 1 data, as well as equal clinical and economic value if the true positive is discovered and validated. However, the same analysis can be easily extended to therapeutic hypotheses of unequal merit as judged by POS or as weighted by clinical or economic value. When applied to two competing hypotheses, the hypothesis with the greater merit will get the larger POC trial, and in the limit of highly unequal merit, all of the resources are devoted to a larger POC trial for the more meritorious hypothesis, mirroring the traditional paradigm. When there are more than two competing trials of unequal merit, the same mathematics can be used to optimize resource allocation across a portfolio of potential POC trials under a fixed budget [12, 13, 39].

The benefit–cost ratio function was designed to require a minimum number of inputs or estimates from the team. For a single program, it requires only estimates of the minimum effect size of clinical interest for the POC endpoint, the minimum effect size of clinical interest for the pivotal endpoint, and the probability that the therapy is actually capable of delivering the latter. If multiple programs are compared, the relative value of each of these indication/therapy combinations needs to be estimated, either in the form of net present value (NPV) or quality adjusted life years (QALY) multiplied by the number of patients who will benefit, or any alternative relative measure of interest to the team. The simplicity of the benefit–cost ratio function allows frequent rapid analyses of complex portfolios and intuitive interpretation of the results. More complex analyses may be difficult to implement in the time before the information becomes out of date, may produce results that are more difficult to intuitively understand, and may not offer enhanced precision due to the difficulty of obtaining accurate values of the many required inputs.

As developed in current references, the benefit–cost ratio is optimized across a portfolio of therapy/indication combinations, each represented by a POC trial. However, this concept can potentially be broadened to a portfolio of therapy–indication combinations and their corresponding primary predictive biomarker hypotheses. This is the core concept for portfolio optimization of an integrated portfolio consisting of both therapeutic and predictive biomarker assets. It is applied in section "Predictive Biomarker Hypotheses: A Global Portfolio View" in this chapter.

The benefit–cost ratio or related utility measures can also be applied within a single indication for optimization of the trial or program design. In the tactics described in section "Tactics" of this chapter, such optimization based on utility functions is applied to the sizing of the Phase 2 trial cohorts, to the decisions regarding further drug and/or biomarker development in Phase 3, and to the optimal design of an adaptive Phase 3 study in the case in which the truth or falsity of the predictive biomarker hypothesis is still uncertain after Phase 2.

Other Fundamental Principles

Two additional principles are fundamental to the approach discussed in Beckman, Clark, and Chen [5]: repeated adaptation throughout the program, and continuous integration of biomarker and clinical information. These principles are illustrated in the tactics described in section "Tactics" below.

Tactics

In this section, we describe four tactics for clinical validation of predictive biomarkers with optimal efficiency. We begin with the efficiency optimized biomarker stratified randomized Phase 2 study, which provides the initial validation test. In this section, we discuss several other tactical considerations relative to therapy-diagnostic co-development in Phase 2. After this Phase 2 study is completed, we face a Go–No Go decision for both the therapy and the predictive biomarker, which is addressed by a decision analysis-guided randomized Phase 2–3 predictive biomarker transition. If the truth or falsity of the predictive biomarker hypothesis is still uncertain after Phase 2, the question of whether to focus on the full population or the predictive biomarker defined subpopulation will have to be further investigated in Phase 3, and in this case we will recommend adaptive-predictive performance based hypothesis prioritization in Phase 3. Finally, this adaptive prioritization is further augmented by maturing Phase 2 data, utilizing the Phase 2+ method.

Efficiency Optimized Biomarker Stratified Randomized Phase 2 Study

There are three options for phase 2 POC studies involving predictive biomarker subgroups. Enrichment designs enroll predominantly or exclusively "biomarker positive" patients. They may be an efficient way to get POC for the therapy, but they do not test the predictive biomarker hypothesis. They are recommended only in the absence of equipoise concerning the biomarker hypothesis. The biomarker strategy design randomizes between two methods of patient allocation between therapies: truly random allocation, or biomarker directed allocation. Finally, the biomarker stratified design stratifies patients into biomarker positive and biomarker negative subgroups, and randomizes to experimental or control therapy within each subgroup, generally in an add-on design, as discussed in section "Core Principles of Predictive Biomarker Evaluation." Thus, there are four groups: biomarker positive experimental, biomarker positive control, biomarker negative experimental, and biomarker negative control. Freidlin, McShane, and Korn [19] have shown that the biomarker stratified design is the most efficient when there is equipoise, since the biomarker strategy design dilutes its power by potentially making the same patient assignment on either arm.

In the biomarker stratified design, two POC hypotheses are being tested at once: one hypothesis concerning the effectiveness of the therapy, and the other hypothesis concerning the relevance of the predictive biomarker classifier. Alternately, this may be formulated as one hypothesis being testing the therapy in the full population and the second hypothesis testing the therapy in the biomarker positive subgroup. A third formulation involves one hypothesis being testing the therapy in the biomarker positive subgroup and the second testing the therapy in the biomarker negative subgroup. As the study is powered for two hypotheses, it will be at least twice the size of the normal Phase 2 study. However, if each subgroup is powered at the Chen–Beckman power, the development efficiency is optimized and the overall study size becomes manageable [5]. The study is considered positive if a therapeutic effect can be demonstrated in either the full population *or* in the biomarker positive subset. Whether investment in a larger Phase 2 study testing a predictive biomarker hypothesis is warranted is a portfolio level optimization question, and is discussed in section "Predictive Biomarker Hypotheses: A Global Portfolio View."

The optimal statistical power occurs when the biomarker positive and negative subgroups are approximately equal in size, and this may require some enrichment. Molecular epidemiology studies should have been undertaken prior to Phase 2 on purchased tissues to determine the prevalence in the population as an aid to study design. In previous work, we have recommended not requiring a biomarker result at enrollment, rather just requiring the availability of tissue for testing [5]. This allows enrollment of patients without waiting for the turnaround time of the predictive biomarker assay. For some patients who are acutely ill with their malignancy and cannot wait to initiate therapy, this makes study participation feasible. This then requires an interim check of the size of biomarker positive and negative groups to determine if prospective enrichment is required later in the study.

However, recent experience indicates that it is important to ensure not only that patients have available tissue but that this tissue is adequate for determining a biomarker result. Depending on the tumor type, inadequate tissue samples can be common. Given the aggressive Chen–Beckman powering, we require near complete ascertainment of biomarker status, and cannot afford to have patients whose tissue proves to be inadequate. This then requires rapid turnaround of the biomarker assay results. If the therapy has multiple indications in which POC can be sought, we recommend testing the biomarker hypothesis in an indication in which the patients can wait for the biomarker result before initiating therapy. Alternatively, if the therapy is intended only for an indication in which immediate treatment is required, patients without an available biomarker result will not be counted in the primary analysis and will need to be replaced. A secondary control analysis will be needed to determine if there is a significant difference in outcome between those with and without available biomarker data. If such a difference is present, it confounds conclusions from the study.

For biomarkers that have a continuous readout, such as gene or protein expression, the definition of the categories "biomarker positive" and "biomarker negative" may not be final until an optimal cutoff value is defined using clinical data. Nonetheless, a provisional cutoff must be set prospectively before clinical data are available to allow prospective testing in the clinical study. In some cases, work with human tissues prior to the study may have defined a bimodal distribution of expression values, and these may have even been associated with distinct biological or pharmacodynamic behavior in preclinical or early clinical work. In such cases, a fundamental cutoff at the intersection point of the two separate modes of the distribution can be chosen [31]. If such information is absent, it may be necessary to set an initial cutoff arbitrarily. In this case, we would suggest setting it at the median value obtained in the Phase 2 study patients, prior to the availability of clinical data. This guarantees equal subgroups for the primary analysis. If the primary analysis is positive, post-hoc refinement of the cutoff against the clinical data would then need to be undertaken. In cases where the cutoff is set at study end using biomarker data (blinded to clinical data), one cannot stratify enrollment by biomarker subgroup, and one must be alert to other confounding factors in the subgroups. Optimization of the cutoff for a continuous biomarker is an iterative process of refinement between biomarker and clinical data. Fridlyand et al. [21] describe further adjustment of the cutoff after a successful Phase 3 based on the original prospective Phase 3 cutoff, a forward-looking approach which does not yet have clear acceptance from health authorities.

In some cases, it may not be possible to choose a primary predictive biomarker hypothesis before the start of phase 2, as fundamental understanding of mechanism of action and relevant biomarkers may lag behind clinical development. In such cases, the primary predictive biomarker hypothesis may be chosen at any time before unblinding of the clinical database, the "prospective-retrospective approach" [34]. The intention to choose a primary predictive biomarker hypothesis before data unblinding should be specified in the protocol. When the primary predictive biomarker hypothesis is chosen, the statistical analysis plan, and possibly the protocol, should be amended to give the study hypothesis in the biomarker subset equal status with the full population as a primary endpoint. Amending the protocol to make this clear, despite its administrative burden, is the most effective approach for making the validity of the analysis plan clear to external stakeholders. When the prospective-retrospective approach is used, stratification of biomarker positive and negative subgroups is not possible, and again one must be alert to other imbalances confounding the subgroup results.

Another reason to use the prospective-retrospective approach is if the predictive biomarker assay is not ready at the start of Phase 2. Minimally, one should have a research use only (RUO) assay available, at the start, but this is not always possible. In that case, again, the samples can be assayed just before the clinical study results are available, with the same disadvantages as in the scenario when the hypothesis is not known.

Full analytical validation of the RUO will result in an in vitro diagnostic (IVD) candidate, which requires only final validation against the clinical phase 3 data to be approved together with the therapy. It is desirable to have the IVD candidate ready before phase 3 to allow submission of an investigational device exemption (IDE) application in that the assay can then be used to select patients in Phase 3 and there is no need for a separate concordance study between RUO and IVD candidate on the Phase 3 samples. Such a concordance study can delay approval and also carries a risk of non-concordance, invalidating the Phase 3 results. Moreover, FDA pre-approval of the Phase 3 study design by Special Protocol Assessment (SPA) is not possible unless an IDE is submitted. However, full validation of the RUO to make an IVD candidate can take 1–2 years, and therefore an important portfolio decision is whether to invest in IVD candidate development prior to POC for the therapy and biomarker to be certain the IVD candidate is ready and does not delay Phase 3 start. Depending on the assay, this investment can range as high as $10–20 M.

Another important consideration is the sample preparation procedure for which the predictive biomarker assay is optimized. Currently, the majority of available patient samples are from the past (archival samples) and are formalin fixed and paraffin-embedded (FFPE). Thus we currently recommend developing assays for use in FFPE to the extent feasible so that the technique is generalizable to the greatest number of patients. However, archival FFPE samples have several disadvantages, including instability of some analytes under these conditions, as well as evolution of the tumor molecular status between tissue acquisition and when the therapeutic is actually applied [38]. In the future, we look forward to increasing practicality of noninvasive methods that allow real time sampling that will monitor the current molecular status of the tumor, not suffer from degradation of analytes over time, and possibly not require fixation [6]. In the future, new fixatives that preserve labile analytes may also enter into widespread use [25].

Finally, we generally recommend a continuous endpoint like progression-free survival (PFS) instead of a discrete endpoint like response rate (tumor shrinkage) which may be uninformative for targeted therapies that have low response rates. With the exception of the neoadjuvant setting, where pathologic complete response is highly correlated with OS in breast cancer [18], PFS is also a better predictor of overall survival (OS) than response rate [15, 36].

Decision Analysis-Guided Phase 2–3 Predictive Biomarker Transition

At the end of the biomarker stratified efficiency optimized randomized Phase 2 study, there are four possible outcomes and decisions, which can be represented on a two dimensional graph with the p values for therapeutic efficacy in biomarker positive and negative patients on the x and y axes respectively (Fig. 10.2).

- Region 1: The therapy is ineffective in both subgroups, resulting in a No-Go decision for further development.
- Region 2: The therapy is effective in the biomarker positive subgroup only. This is a POC for both the therapy and the predictive biomarker hypothesis, and results in a decision to proceed with Phase 3 in the biomarker positive subgroup only.
- Region 3: The therapy is equally effective in both biomarker positive and negative subgroups. This is a POC for the therapy, and a rejection of the predictive biomarker hypothesis. It results in a decision to proceed to Phase 3 in an unselected population.
- Region 4: The therapy is effective in the biomarker positive population and there is a trend towards effectiveness, although less, in the biomarker negative population. There is a POC for the therapy, but the status of the biomarker hypothesis is uncertain. In this case we recommend proceeding to further adaptation in Phase 3, using adaptive predictive performance based hypothesis prioritization (section "Adaptive, Predictive Performance-Based Hypothesis Prioritization in Phase 3") in conjunction with the Phase 2+ method (section "The Phase 2+ Method for Allowing Phase 2 Data to Influence Adaptation Within Phase 3").

In order to optimally draw the borders between the regions of the graph, a decision analysis is undertaken [5]. Each possible choice of phase 3 development path will result in several possible outcomes, with the probability of each determined by the Phase 2 data. For example, if a choice is made to do an enriched Phase 3 study in the biomarker positive population only and the predictive biomarker hypothesis is true, the outcome will be an approval in an optimal population, possibly with a lower cost trial if there is a large effect size in the biomarker positive group (see section "Predictive Biomarker Hypotheses: A Global Portfolio View"). But if the same choice is made and the predictive biomarker hypothesis is false, there will be an opportunity cost in that the approval population will have been unnecessarily narrowed. Similarly, if Phase 3 proceeds in the full population and the predictive biomarker hypothesis is false, there will be an approval in the optimal population. But if the same choice is made and the biomarker hypothesis is true, there may be a falsely negative Phase 3 due to dilution of the effect by biomarker negative patients. Once utility values are assigned to each of these possible outcomes, the optimal borders, which optimize utility subject to a predetermined maximum phase 2 false positive rate, can be drawn.

We have previously advocated that the utility values be set by discussion within the expert team that is managing the development of the therapy. However, it would

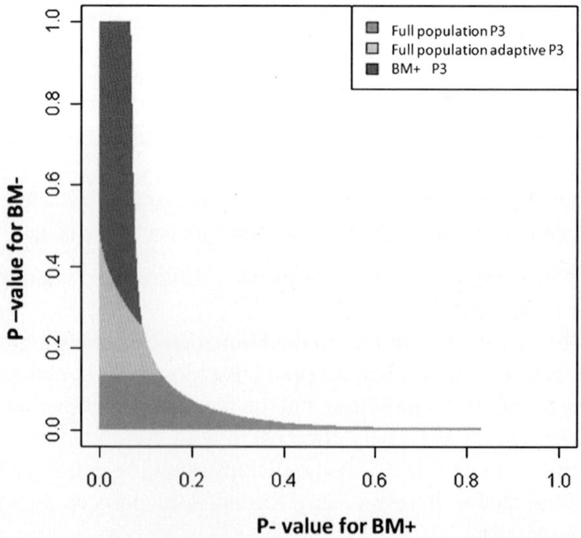

Fig. 10.2 Two dimensional decision graph for decision analysis guided Phase 2- Phase 3 predictive biomarker transition. The *Y*-axis shows the one-sided *p* value for clinical benefit for "biomarker negative" (BM–) patients in the PoC study, the *X*-axis shows the one-sided *p* value for clinical benefit for "biomarker positive" (BM+) patients in the PoC study. Clinical benefit is greater the closer the point is to the origin. The graph has four regions corresponding to the four possible decisions: (R1) *upper-right* (*white*): No Go; (R2) *upper-left* (*dark grey*): go to biomarker positive only Phase 3; (R3) *lower-left* (*medium grey*): Go to traditional Phase 3 in full population; (R4) *middle-left* (*light grey*): go to biomarker adaptive Phase 3 in full population (adaptive predictive performance-based hypothesis prioritization);. The graph is calculated using a Bayesian decision theoretic model [7] using three possible assumed scenarios and estimates of their probabilities: (H0) a "null" hypothesis where the experimental drug regimen provides no benefit over standard of care; (H1) an "all comers" hypothesis where the drug is effective but the biomarker does not define a difference in patient benefit for the experimental regimen over control; and, (H2) a "biomarker only" hypothesis where the biomarker clearly defines if patients benefit from the experimental regimen. Based on the outcome of a randomized Phase 2 trial with stratification for patients according to the biomarker of interest, the initial team-estimated probabilities assigned to the three scenarios (prior probability estimates) are updated using Bayes' theorem to create posterior probability estimates. Using these posterior probability estimates and team estimates of relative utilities of different consequences which may result from the choices in the three scenarios, the resulting regions are selected to maximize the expected utility. This optimization is performed under the constraint that the chance of falsely going to Phase 3 when the drug is ineffective (the null hypothesis H0) will be equal to or less than the one-sided false positive rate α. The graph shows a case assuming equal numbers of biomarker positive and negative patients. The boundaries of the regions would differ depending on the ratio of these two populations and other factors. Thus, the graph is only an example, and the exact boundaries need to be calculated for each specific case. Adapted from Beckman, Clark, and Chen [5]

also be of interest to study how these utility values would be set by patients, insurers, or developers of diagnostic assays, and how this may affect the optimal borders between regions of the graph.

For continuous biomarkers, it will be useful to draw one of these graphs for the initial arbitrary cutoff value between biomarker positive and negative, and another for the refined cutoff determined at the end of the study using clinical data. If the decision differs between the two graphs, it may be helpful to look at a sensitivity analysis of the optimal decision as a function of the cutoff.

Adaptive, Predictive Performance-Based Hypothesis Prioritization in Phase 3

If the Phase 2 data are in region 4 of the graph in Fig. 10.2, ambiguity remains about the truth of the predictive biomarker hypothesis. Based on the underlying principles of adaptation and continuous integration of biomarker and clinical results, we continue the investigation of the biomarker hypothesis in an adaptive Phase 3 design [14]. This design allows simultaneous testing of two hypotheses:

1. Hypothesis 1: The therapy works in the full population.
2. Hypothesis 2: The therapy works in the subpopulation defined by the predictive biomarker hypothesis.

In order to test both hypotheses, we need to enroll the full population and test hypothesis 1 in the full population while *simultaneously* testing hypothesis 2 in the subset. We note that other designs are available which involve sequential testing, where one of the hypotheses is tested first, and the second tested only if the first test is positive. Sequential testing approaches are efficient if it is more likely that the first hypothesis tested will be positive as compared to the second one. But in this case the truth of the full population hypothesis and the truth of the predictive biomarker hypothesis are both uncertain, so the proper order for sequential testing is unknown.

We will thus test both hypotheses simultaneously, and the trial will be defined as a positive trial if either of the hypothesis tests is successful. Because there are two ways for the trial to be judged positive, care must be exercised to keep the false positive rate (alpha) at 5 % or less, as stipulated by national health authorities. This is done by "splitting" the 5 % positive rate, into lower rates for each of the two hypotheses. The total false positive rate can then be computed and set to 5 %. Because hypothesis 2 involves a subset of the sample from hypothesis 1, there is some degree of correlation between the hypotheses, such that the total false positive rate will be less than the sum of the two individual false positive rates; but in any event the total false positive rate can be fixed at 5 %.

In keeping with our fundamental principles, we would like to optimize the efficiency of this trial in a data-driven manner. The solution is to split alpha in a way which optimizes the expected power of the study based on clinical data. The better the predictive biomarker hypothesis has been doing in predicting the clinical data, the larger fraction of the alpha is devoted to testing Hypothesis 2. In effect, in the setting where we do not know enough to place a full bet on either Hypothesis 1 or

Hypothesis 2, the methodology computes the optimal way to hedge our bets [14]. Net study power will be superior to methods where a fixed alpha split is applied to test two hypotheses [20]. Methods that preserve alpha using the combination test have also been developed [9, 22].

The adaptation in which alpha split is chosen will be at an interim analysis point in Phase 3. The clinical data used can either be from Phase 3 to that point (which involves a statistical penalty for using data from within the study for a within study adaptation) or external to the study (i.e., from Phase 2; see section "The Phase 2+ Method for Allowing Phase 2 Data to Influence Adaptation Within Phase 3"). The rules for determining the alpha split can be determined prospectively and given to an independent Data Monitoring Committee (DMC) for automatic application prior to the availability of final study results. In addition to the prospective rule setting and independent application, the procedure does not affect patient selection or management, and strictly controls the false-positive rate alpha. As such, there should be no concerns about this methodology from national health authorities or local ethics review boards.

The Phase 2+ Method for Allowing Phase 2 Data to Influence Adaptation Within Phase 3

In keeping with the core principles stated in section "Core Principles of Predictive Biomarker Evaluation," we would like to continuously adapt as we proceed through development. However, real time adaptation in oncology is hampered by the fact that the clinical endpoint of greatest interest, OS, is often measured only with a significant delay. Thus, we frequently adapt based on short term endpoints that have imperfect correlation with OS, such as PFS.

Typically Go–No Go decisions are made based on Phase 2 PFS data because it is undesirable to slow development while waiting for mature OS data. Yet the primary endpoint of Phase 3 will be OS.

If there is an adaptation within Phase 3, it will be performed at interim analysis, typically based on interim Phase 3 data. At this time, the OS data will largely be immature, so again the adaptation will be based on PFS, despite its imperfect correlation with the primary endpoint OS. In addition to the imperfect correlation with the primary endpoint, this paradigm suffers from the possible inflation of the Phase 3 false positive rate, which requires statistical penalties to be applied to assure compliance with maximum allowable false positive rates for Phase 3 pivotal studies.

An alternative approach, the Phase 2+ method, has been proposed ([5, 14]; Fig. 10.3) in which maturing Phase 2 data can be used to direct a Phase 3 adaptation. Typically, there will be a point during the Phase 3 study, before Phase 3 results are unblinded, at which Phase 2 OS data are mature. By using Phase 2 OS data which is external to the Phase 3 study, there is no issue of inflation of the false positive rate of Phase 3 due to the interim adjustment of the alpha split. (Nonetheless, one may optionally choose to use OS data from Phase 3 at interim with appropriate penalties, as illustrated in Fig. 10.3). Most significantly, the adaption is on OS data, the same

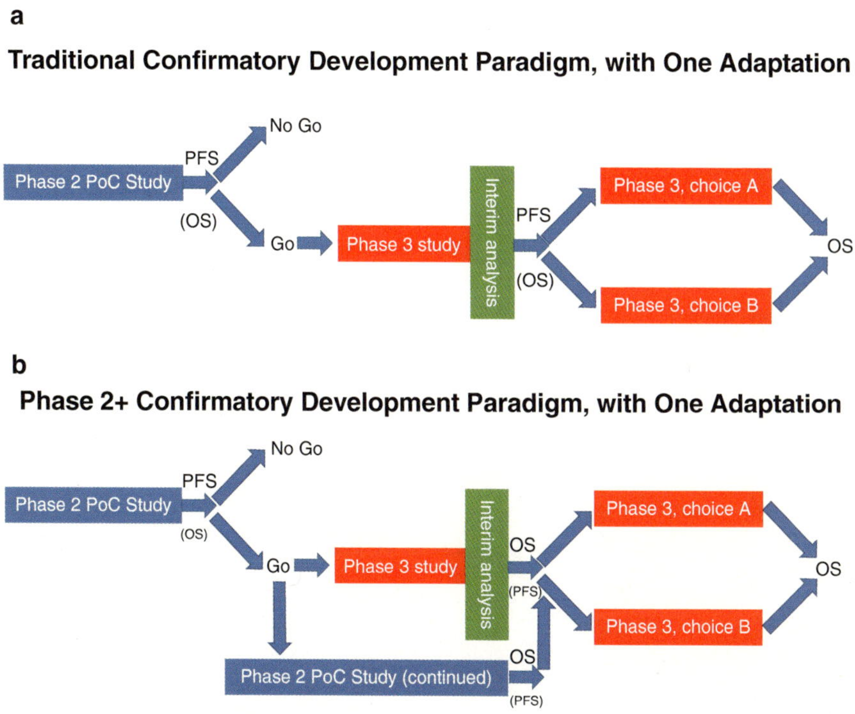

Fig. 10.3 Traditional confirmatory phase development compared to the Phase 2+ method. Phase 2 PoC studies (*blue*) are shown on the *left*, Phase 3 approval studies (*red*) on the *right*. Primary endpoints are progression free survival (PFS) for Phase 2, overall survival (OS) for Phase 3. In the *traditional paradigm with one adaptation* (**a**), the Phase 2 PoC study is completed, and a Go–No Go decision based on PFS (with or without some consideration of immature OS data, in parentheses) determines whether a Phase 3 approval study is done. The Phase 3 study has an interim analysis (olive), based again primarily on PFS (with or without consideration of immature OS data from Phase 3, in parentheses), and an adaptation occurs based on this interim analysis (represented by two different Phase 3 bars). In the *Phase 2+ paradigm with one adaptation* (**b**), multiple analyses of the Phase 2 study occur, and the Phase 2 and 3 studies functionally overlap. The first analysis of the Phase 2 PoC trial results in a Go–No Go decision for Phase 3, as in (**a**). But unlike (**a**), Phase 2 data are utilized for decision making up to and including the Phase 3 interim analysis. The Phase 3 interim analysis includes maturing OS data from Phase 2, possibly PFS data up to the first Phase 2 analysis, and possibly immature OS and PFS data from Phase 3. These multiple data sources, particularly the maturing Phase 2 OS data, contribute to a more robust adaptation at the Phase 3 interim analysis. An alternative option, using the Phase 2 OS data only, provides a very straightforward approach without the need for statistical penalties for using the Phase 3 data for adaptation, and without needing to adjust for the imperfect correlation between PFS and OS. Adapted from Beckman, Clark, and Chen [5]

endpoint that will be pivotal for Phase 3. Previous development paradigms have not harvested the full value of this maturing Phase 2 OS data. In the Phase 2+ method, additional value results from considering the program as an integrated whole rather than as a collection of isolated individual studies.

In order to maximize the value of this approach, Phase 2 OS must correlate well with Phase 3 OS, and therefore the Phase 2 POC study must be as closely identical to the Phase 3 pivotal study as possible. In particular study populations, inclusion/exclusion criteria, and experimental and control arms should ideally be identical between the Phase 2 and 3 studies. The investigative sites for Phase 2 should be used again in Phase 3, with other similar sites added due to the larger size of Phase 3.

In anticipation of the use of this technique, the Phase 2 study should prespecify several analysis timepoints:

1. Phase 2 primary analysis for PFS and Go–No Go decision. Conducted when PFS data are mature.
2. Final Phase 2 OS analysis conducted when OS data are mature.
3. Phase 2+ OS analysis to support adaptation in Phase 3. Conducted during Phase 3 sometime before data unblinding, as close to the maturity of Phase 2 OS data as possible.

Predictive Biomarker Hypotheses: A Global Portfolio View

Predictive biomarker hypotheses, like candidate therapies, are uncertain assets with associated value and risk. Moreover, just as with candidate therapies, investment is required to even assess the value of a predictive biomarker hypothesis. Finally, just as each candidate therapy is represented by a lead compound and backups, we have defined a primary predictive biomarker hypothesis which has a role analogous to a lead compound. Just as a backup compound may be substituted in some instances in the event of failure of a lead, an exploratory biomarker hypothesis may become primary in the event of failure of the primary predictive biomarker hypothesis. Investigation of either a backup candidate therapy or of a backup predictive biomarker hypothesis will require an additional iteration through development.

Thus the generalized portfolio problem should not be limited to prioritization and resource allocation of candidate therapies, but rather should encompass prioritization and resource allocation among the integrated portfolio of candidate therapies and predictive biomarker hypotheses. For example, one might have a portfolio of ten lead candidate therapies, each with its associated backups, primary predictive biomarker hypothesis, and exploratory biomarker hypotheses. After portfolio optimization in the face of budgetary constraints, resources may be focused on seven of these candidate therapies, with further investment in associated predictive biomarkers for only 5. As another example, a given candidate therapeutic may have sufficient funds for POC testing in five indications of interest without testing its associated predictive biomarker hypothesis. But, if additional moneys are invested for testing the predictive biomarker hypothesis in each indication, the total available funds may be only enough to test three indications. Which plan is more efficient?

We have discussed a core principle of optimizing development efficiency by maximizing a risk adjusted benefit–cost ratio (BCR) across an individual POC program, a collection of POC programs of equal value, or a portfolio of POC programs of unequal value (section "Core Principle 2: Optimizing Development Efficiency," this chapter; [39]). In order to develop the BCR for therapies, we defined the benefits and costs associated with late development of therapies, and risks and probabilities that these benefits and costs would be realized.

In the sections below, we will define the benefit, costs, risks, and probabilities of success for predictive biomarkers, within the context of the paradigm for integration of predictive biomarker hypotheses into individual development programs outlined in sections "Core Principles of Predictive Biomarker Evaluation" and "Tactics." This will provide a qualitative foundation for understanding the integrated portfolio optimization problem. Although explicit mathematical formulations are beyond the scope of this chapter, they can in principle be readily developed based on these concepts.

The Risk-Adjusted Value of a Predictive Biomarker Hypothesis

A predictive biomarker hypothesis lacks inherent value in the absence of a candidate therapy. It derives its value from its effect in decreasing the risk and/or increasing the value associated with a candidate therapy or therapies in one or more indications. For example, a predictive biomarker may decrease the risk of a candidate therapy by increasing the probability that the therapy will successfully deliver a minimally clinically significant result compatible with registration, without actually increasing the effect size substantially beyond that level. A higher value would not be realized for that individual therapy, but a portfolio with lower average risk of failure could be delivered at a lower cost. In contrast, some predictive biomarkers may lead to increased clinical effect sizes beyond those minimally required for registration. Such biomarkers would both lower the risk of development, and increase the value of candidate therapies. This latter class of biomarkers is potentially more valuable when available, but given the very high genetic complexity and instability of cancer [3] caution should be exercised in assuming that the majority of predictive biomarkers will in fact deliver very large clinical effect sizes.

As with the BCR for candidate therapies, we desire to define a simple function with a small number of intuitively understandable inputs that will encompass the value of a predictive biomarker hypothesis.

We propose the following criteria as determinants of the value of a predictive biomarker hypothesis:

1. Level of evidence within the indication: what is the probability the predictive biomarker hypothesis is true in the indication in which it is being tested? Prior to a phase 2 study, evidence pertinent to this question will come from the various preclinical, mechanism of action, and early clinical studies outlined in section "Core Principle 1: Clinical Validation of Predictive Biomarkers."
2. Impact on therapeutic development risk: if the predictive biomarker hypothesis is true, to what degree will it increase the probability that the therapy can achieve minimally clinically significant effects in the new subpopulation defined by the hypothesis? This amounts to reduction of development risk, and is perhaps the key benefit of successful predictive biomarker hypotheses. The same sources of pertinent evidence as in 1 above must be considered when estimating this.

3. Impact on therapeutic value: if we define therapeutic value as the number of patients benefitting times the benefit (the latter measured in QALY for example), the predictive biomarker hypothesis, if true, will result in a decrease in the number of patients benefiting but often (not always) an increase in the average benefit per patient. The total therapeutic value is the product of the number of patients and the average benefit per patient, and may increase or decrease. The change in average therapeutic benefit in the biomarker defined subpopulations may again be estimated from preclinical and early clinical studies, and the change in population size may be estimated from molecular epidemiology studies estimating biomarker prevalence in human tissues.
   (a) Probability of a true positive test of the biomarker hypothesis: Based on the Phase 2–3 development paradigm described in sections "Core Principles of Predictive Biomarker Evaluation" and "Tactics," and the estimated average change in patient benefit in the predictive biomarker defined subpopulation, it should be possible by simulation to determine the probability that a true predictive biomarker hypothesis will actually be detected as such by the development program (as opposed to resulting in a false negative). In order for a predictive biomarker hypothesis to benefit a program, it must be true *and* its truth must be detected by the development program (that is the test of the predictive biomarker hypothesis must lead to a true positive).
   (b) Probability of a false positive test of the predictive biomarker hypothesis: Based on the Phase 2–3 program design and the probability that the predictive biomarker hypothesis is actually true or false, it should be possible to simulate the overall probability that a false positive test of the biomarker hypothesis occurs. A false positive test of a biomarker hypothesis decreases the value of a candidate therapy in that the population benefitting is unnecessarily narrowed.
   (c) Probability of a true or false negative test of the predictive biomarker hypothesis: These events do not change the value of a therapeutic candidate and its indication relative to the value without a biomarker hypothesis. Therefore, these individual probabilities need not be estimated.
4. Generalizability: what is the probability that the predictive biomarker hypothesis will benefit multiple indications? The strategic value of a predictive biomarker hypothesis is greater if it can be generalized to other indications. The value could be immense if the predictive biomarker hypothesis benefits multiple indications. In such a case, testing the predictive biomarker hypothesis may be a higher priority than testing any single therapeutic hypothesis. However, examples exist where a predictive biomarker hypothesis did not generalize across indications. For example, the BRAF V600E mutation predicts efficacy in melanoma therapy [10] but did not generalize to colorectal cancer [28]. Often failure to generalize is associated with tissue specific feedback loops which generate resistance to targeted therapies. Preclinical pharmacology data on molecularly characterized models, as well as mechanism of action studies, may help in projecting the degree to which a predictive biomarker hypothesis will generalize.
   (a) Information borrowing: what is the likelihood that the predictive biomarker hypothesis, once assessed as true in one or two indications, can be generalized to others without further testing? If the predictive biomarker hypothesis is

shown to be true in both lung cancer and head and neck cancer, is there a need to test it in other indications such as breast and colorectal cancers? This factor affects the cost (not the value) of biomarker development, but is mentioned here because of its relationship to generalizability.

The value of any individual indication for a candidate therapy in the absence of a predictive biomarker hypothesis is the product of three numbers: the probability that the therapy is actually effective in the indication, the probability that the development plan will detect this effectiveness, and the value of the registered indication in the event of a true positive outcome (measured either in total patient benefit or net present value). The value of a portfolio of indications for therapies is the sum of each of these products.

When the decision is made to evaluate a predictive biomarker hypothesis, each relevant indication has an incremental value added, which may be positive or negative. A cost is also added, and this may decrease the number of indications that can be tested within the budget.

In the event of a true positive test of the predictive biomarker hypothesis, the new value of the indication will reflect an increased probability that the drug actually works in the indication (in the new subpopulation defined by the predictive biomarker), a possibly increased probability that this effectiveness will be detected by the clinical studies (if a larger clinical effect size is anticipated), a reduced number of patients benefitting, and possibly an increased average benefit per patient. The incremental value of a true positive predictive biomarker hypothesis with respect to a given therapy and indication is the new indication value in the presence of the predictive biomarker minus the value in the unselected population without a predictive biomarker.

In the event of a false positive biomarker test, the new value of the indication will reflect only decreased population size. The incremental value, negative under these circumstances, is again the new indication value in the presence of a false positive biomarker hypothesis minus the indication value in the unselected population without a predictive biomarker.

The total risk adjusted value of a predictive biomarker hypothesis in a given indication is the sum of its incremental values in the presence of a true positive or false positive outcome of the test of the predictive biomarker hypothesis, weighted by the probability of occurrence of a true or false positive predictive biomarker hypothesis respectively.

The total risk adjusted value of a predictive biomarker hypothesis is the sum of its risk adjusted value across all indications over which it is expected to generalize.

The Risk-Adjusted Cost of Testing a Predictive Biomarker Hypothesis

In this section, we will first discuss the risk-adjusted cost of testing a predictive biomarker hypothesis in a single indication, and then consider the risk-adjusted cost of developing a predictive biomarker across multiple applicable indications.

The following criteria determine the risk-adjusted cost of testing a predictive biomarker hypothesis in a single indication:

1. The increased cost of doing a stratified randomized Phase 2 study with two biomarker subgroups as opposed to a Phase 2 study in the unselected population, consisting of:
   (a) The cost of increasing the sample size.
   (b) The cost of increased screening required for enrichment of the less prevalent subgroup. This can be estimated based on the known prevalence of the biomarker from studies in human tissues.
   (c) The cost of increased screening due to patients with inadequate tissue. This can be estimated based on the tissue requirements of the primary biomarker assay and historical data on tissue availability in the indication. The primary biomarker assay should be the first priority for available tissue after the primary diagnosis.
   (d) The cost of tissue management and conducting biomarker assays.
   (e) Increased costs associated with a likely longer duration of study.
2. The cost of developing a companion diagnostic assay at risk.
3. The risk adjusted savings from a smaller Phase 3 study if the following conditions are met:
   (a) A phase 2 result in region 1 of Fig. 10.2, resulting in a Phase 3 study in biomarker positive patients only.
   (b) An increased clinical benefit effect size in biomarker positive patients in Phase 2 that results in the possibility of designing a smaller Phase 3 study because a larger clinical benefit can be assumed. Before taking the risk of reducing the sample size in Phase 3, the development team will need to consider not just the point estimate of the clinical benefit effect size, but also its uncertainty.

The risk adjusted savings from this mechanism can be estimated by simulations given the Phase 2 study design, the decision criteria determining region 1 in Fig. 10.2, the estimated probability that the predictive biomarker hypothesis is true, and the estimated clinical benefit effect size in the biomarker positive population in the event of a true predictive hypothesis.

The cost of developing and utilizing a predictive biomarker hypothesis fully across all relevant indications involves the sum of the costs of developing the biomarker in each relevant indication unless there is a plan to assume generalizability of the predictive biomarker hypothesis across indications. We believe it is reasonable in most cases to assume generalizability despite the risk involved. For example, if the predictive biomarker hypothesis fails in the first indication, one might not test it in further indications. If the predictive biomarker succeeds in the first two indications, one might assume its truth for future relevant indications and develop the therapy in predictive biomarker defined subpopulations only in future indications.

Such plans seem intuitively sensible and will reduce costs, but in estimating their efficiency one must bear in mind that any assumption of generalizability increases the risk of false positives and negatives with respect to the biomarker hypothesis. It is unclear what data can be used to quantify the risk of false positives and negatives under these circumstances.

Another cost of developing a predictive biomarker hypothesis in multiple indications may be adapting the companion diagnostic assay to the different indications. While some assays may be readily adaptable, others will require tissue specific versions at a reduced but non-zero cost. For example, gene expression assays must be normalized to "housekeeping genes" with constant expression. The choice of housekeeping genes may be tissue specific.

Portfolio Analysis Involving Predictive Biomarker and Candidate Therapy Assets

From the above, the complexity of portfolio analysis involving both predictive biomarker hypotheses and therapeutic candidates is apparent. Calculating false positive and negative rates for testing the predictive biomarker hypothesis from the complex adaptive Phase 2–3 development paradigm likely requires simulations. Predicting whether a successful predictive biomarker hypothesis will only increase the probability of achieving a minimally significant clinical benefit effect size or will also increase the clinical benefit effect size may be challenging. Estimating the degree of generalizability of the predictive biomarker hypothesis across indications may also be difficult. Multiple possible biomarker development plans result from different assumptions about generalizability of results. Individual biomarker assay characteristics will determine the incremental cost of developing a companion diagnostic across multiple indications.

We recommend that each therapeutic program and associated predictive biomarker hypothesis be considered in three scenarios. Within each scenario, the optimal allocation of investment between indications can be determined using previously published methods [12, 13, 39].

1. With the predictive biomarker hypothesis developed in each relevant indication.
2. With a predictive biomarker hypothesis developed in a small subset of indications and assuming generalizability of the results to other indications.
3. Without a predictive biomarker hypothesis.

A comparison of the benefit–cost ratio in these three scenarios should give guidance as to which plan is optimal for a given therapy candidate. This can be followed by a ranking across therapeutic programs of the optimal development scenarios by the benefit–cost ratio and hence overall prioritization within a fixed budget for POC studies and biomarker development. Additional "what if" scenarios can be addressed ad hoc by computing benefit–cost ratios for alternative portfolio strategies.

Summary and Future Directions in Personalized Therapy of Cancer

This chapter has begun by outlining an approach to integration of predictive biomarker hypotheses into individual oncology clinical development programs. The core principles include clinical validation of predictive biomarker hypotheses, and optimization of programs based on maximizing a benefit–cost ratio. Corollaries of these principles include the need to include biomarker negative patients in late development, and prioritization of a single primary predictive hypothesis for formal statistical testing in phase 2. The resulting program is adaptive and data-driven, and demands value from predictive biomarkers. This may result in the rejection of some putative predictive biomarker hypotheses, but ultimately will result in greater value.

The benefit–cost ratio has been previously developed not only for individual program optimization but also for optimization of resource allocation across a portfolio of candidate therapies and their indications. The chapter begins to extend that work by considering optimal management of an integrated portfolio involving two asset classes: candidate therapeutics and putative predictive biomarker hypotheses. This again involves maximizing a benefit–cost ratio that is designed to be as simple and intuitive as possible. However, the complexity of optimizing a portfolio with two mutually interacting asset classes is unavoidably greater than optimization across therapeutic candidates alone. Nonetheless, portfolio analysis is valuable to optimally manage risk given that both putative biomarker hypotheses and candidate therapies are uncertain.

How will personalized cancer therapy change in the future? The current approach to personalized medicine involves matching patients to optimal therapies by their tumors' molecular characteristics. This is done based on average molecular properties of bulk samples at diagnosis and (when possible) relapse.

However, tumors are genetically unstable, because this is the most efficient way for them to evolve [2–4]. As a result, there will be subclones whose molecular properties are not detected in bulk samples. These subclones can lead to preexisting resistance to targeted therapy. Rapid evolution of acquired resistance due to genetic instability is another source of resistance.

Current personalized medicine does not fully consider this intra-tumoral heterogeneity and evolutionary dynamics. More complex nonstandard personalized strategies that explicitly consider subclonal structure (including the risk of rare cells below the detection limit) and evolutionary dynamics (including the risk of predicted future states) have the potential to dramatically improve patient outcomes, as determined by extensive simulation studies [6]. Clinical translation of these ideas will require improved noninvasive access to tumors and single-cell analysis techniques as well as deeper understanding of molecular mechanisms of tumor evolution and resistance.

Nonstandard personalized medicine strategies will in the future think several moves ahead, like a chess master, as opposed to the current approach which matches patients to therapies one therapeutic maneuver at a time. However, to play chess one must first know the pieces and how they move. The elucidation and validation of predictive biomarker hypotheses for available therapies continues to define the very large number of pieces and moves in the vitally important game of personalized cancer therapy.

Acknowledgements The authors wish to thank Jason Clark for contributing the decision analysis guided Phase 2-Phase 3 predictive biomarker transition, and Donald Bergstrom, Daniel Freeman, Robert Phillips, Richard M. Simon, and Linda Sun for helpful discussions.

References

1. Amado RG, Wolf M, Peeters M et al (2008) Wild-type KRAS is required for panitumumab efficacy in patients with metastatic colorectal cancer. J Clin Oncol 26:1626–1634
2. Beckman RA (2009) Mutator mutations enhance tumorigenic efficiency across fitness landscapes. PLoS One 4:e5860
3. Beckman RA (2010) Efficiency of carcinogenesis: in the mutator phenotype inevitable? Semin Cancer Biol 20:340–352
4. Beckman RA, Loeb LA (2006) Efficiency of carcinogenesis with and without a mutator mutation. Proc Natl Acad Sci U S A 103:14410–14415
5. Beckman RA, Clark J, Chen C (2011) Integrating predictive biomarkers and classifiers into oncology clinical development programs. Nat Rev Drug Discov 10:735–748
6. Beckman RA, Schemmann GS, Yeang CH (2012) Impact of genetic dynamics and single-cell heterogeneity on development on nonstandard personalized medicine strategies for cancer. Proc Natl Acad Sci U S A 109:14586–14591
7. Berger J (1985) Statistical decision theory and bayesian analysis, 2nd edn. Springer, New York
8. Bokemeyer C, Bondarenko I, Hartmann JT et al (2008) KRAS status and efficacy of first-line treatment of patients with metastatic colorectal cancer (mCRC) with FOLFOX with or without cetuximab: the OPUS experience. J Clin Oncol 26(Suppl. 15S), abstract
9. Brannath W, Zuber E, Branson M et al (2009) Confirmatory adaptive designs with Bayesian decision tools for a targeted therapy in oncology. Stat Med 28:1445–1463
10. Chapman PD, Hauschild A, Robert C et al (2011) Improved survival with vemurafenib in melanoma with BRAF V600E mutation. N Engl J Med 364:2507–2516
11. Chen C, Beckman RA (2007) Optimal cost effective designs of proof of concept trials and associated Go-No Go decisions. Proceedings of the American Statistical Association, Biometrics Section
12. Chen C, Beckman RA (2009) Optimal cost-effective Go-No Go decisions in late stage oncology drug development. Stat Biopharm Res 1:159–169
13. Chen C, Beckman RA (2009) Optimal cost-effective Phase II proof of concept and associated Go-No Go decisions. J Biopharm Stat 1:424–436
14. Chen C, Beckman RA (2009) Hypothesis testing in a confirmatory Phase III trial with a possible subset effect. Stat Biopharm Res 1:431–440
15. Chen C, Sun L, Chih C (2013) Evaluation of early efficacy endpoints for proof-of-concept trials. J Biopharm Stat 23:413–424
16. Cobleigh MA, Vogel CL, Tripathy D et al (1999) Multinational study of the efficacy and safety of humanized anti-HER2 monoclonal antibody in women who have HER2-overexpressing metastatic breast cancer that has progressed after chemotherapy for metastatic disease. J Clin Oncol 17:2639–2648
17. Dalton WS, Friend SH (2006) Cancer biomarkers—an invitation to the table. Science 312:1165–1168

18. Esserman LJ, Berry DA, DeMichele A et al (2012) Pathologic complete response predicts recurrence-free survival more effectively by cancer subset: results from the I-SPY 1 TRIAL—CALGB 15007/150012, ACRIN 6657. J Clin Oncol 30:3242–3249
19. Freidlin B, McShane L, Korn EL (2010) Randomized clinical trials with biomarkers: design issues. J Natl Cancer Inst 102:152–160
20. Freidlin B, Simon RM (2005) Adaptive signature design: an adaptive clinical trial design for generating and prospectively testing a gene expression signature for sensitive patients. Clin Cancer Res 11:7872–7878
21. Fridlyand J, Simon RM, Walrath JC et al (2013) Considerations for the successful co-development of targeted therapies and companion diagnostics. Nat Rev Drug Discov 12:743–755
22. Jenkins M, Stone A, Jennison C (2011) An adaptive seamless phase II/III design for oncology trials with subpopulation selection using correlated survival endpoints. Pharm Stat 10:347–356
23. Kelloff GJ, Sigman CC (2012) Cancer biomarkers: selecting the right drug for the right patient. Nat Rev Drug Discov 11:201–214
24. Lievre A, Bachet JB, Le Corre D et al (2006) KRAS mutation status is predictive of response to cetuximab therapy in colorectal cancer. Cancer Res 66:3992–3995
25. Mueller C, Edmiston KH, Carpenter C et al (2011) One-step preservation of phosphoproteins and tissue morphology at room temperature for diagnostic and research specimens. PLoS One 6:e23780
26. Paez JG, Janne PA, Lee JC et al (2004) EGFR mutations in lung cancer: correlation with clinical response to gefitinib therapy. Science 30:1497–1500
27. Pao W, Miller V, Zakowski M et al (2004) EGF receptor gene mutations are common in lung cancers from "never smokers" and are associated with sensitivity of tumors to gefitinib and erlotinib. Proc Natl Acad Sci U S A 101:13306–13311
28. Prahallad A, Sun C, Huang S et al (2012) Unresponsiveness of colon cancer to BRAF(V600E) inhibition through feedback activation of EGFR. Nature 483:100–103
29. Ransohoff DF, Gourlay ML (2010) Sources of bias in specimens for research about molecular markers for cancer. J Clin Oncol 28:698–704
30. Ratain MJ, Glassman RH (2007) Biomarkers in phase I oncology trials: signal, noise, or expensive distraction? Clin Cancer Res 13:6545–6548
31. Shames DS, Carbon J, Walter K et al (2013) High heregulin is associated with activated HER3 and may define an actionable biomarker in patients with squamous cell carcinoma of the head and neck. PLoS One 8:e56765
32. Shaw AT, Kim DW, Nakagawa K et al (2013) Crizotinib versus chemotherapy in advanced alk-positive lung cancer. N Engl J Med 368:2385–2394
33. Shepherd FA, Pereira JR, Ciuleanu T et al (2005) Erlotinib in previoously treated non-small-cell lung cancer. N Engl J Med 353:123–132
34. Simon RM, Paik S, Hayes DF (2009) Use of archived specimens in evaluation of prognostic and predictive biomarkers. J Natl Cancer Inst 101:1446–1452
35. Slamon DJ, Leyland-Jones B, Shak S et al (2001) Use of chemotherapy plus a monoclonal antibody against HER2 for metastatic breast cancer that overexpresses HER2. N Engl J Med 344:783–792
36. Tang PA, Bentsen SM, Chen EX, Siu LL (2007) Surrogate endpoints for median overall survival in metastatic colorectal cancer: literature-based analysis from 39 randomized controlled trials of first-line chemotherapy. J Clin Oncol 25:4562–4568
37. Van Cutsem E, Lang I, D'haens G et al (2008) KRAS status and efficacy in the first-line treatment of patients with metastatic colorectal cancer (mCRC) treated with FOLFIRI with or without cetuximab: the CRYSTAL experience. J Clin Oncol 26(Suppl. 15S), abstract
38. Yan L, Beckman RA (2005) Pharmacogenetics and pharmacogenomics in oncology therapeutic antibody development. Biotechniques 39:565–568
39. Chen C, Beckman RA, Sun LZ Maximizing Return on Investment in Phase II Proof-of-Concept Trials. The bibliographic information for this volume is in the proofs: Z. Antonijevic (ed.), *Optimization of Pharmaceutical R&D Programs and Portfolios: Design and Investment Strategy*, DOI 10.1007/978-3-319-09075-7_10.

Chapter 11
Dynamically Optimizing Budget Allocation for Phase 3 Drug Development Portfolios Incorporating Uncertainty in the Pipeline

Nitin R. Patel and Suresh Ankolekar

Introduction

In this chapter we describe a model for optimizing allocation of a budget that has been earmarked for Phase 3 (Ph3) development of drugs in a pipeline over a planning horizon. We consider drugs that are ready to enter Ph3 trials as well as drugs that are at earlier stages of testing but are scheduled to require funding for Ph3 trials over the planning horizon. We focus on Ph 3 trials as they are very expensive and because in most large pharmaceutical companies Ph3 budget allocation follows a process that is separate from funding of earlier stages of drug development. (In section "Summary" we indicate how our model can be modified to include both Phase2 (Ph2) and Ph3 budget allocation.)

In chapter "Challenges of Portfolio Management in Pharmaceutical Development" Charles Persinger describes the challenge of applying the traditional portfolio management approach in the "Big Pharma" setting of drug development. He identifies two key elements that constitute the challenge. The first is the fact that at any point in time when we are considering budget allocation to drugs that are ready for Ph3 trials, several compounds in the clinical research pipeline are typically at earlier stages of ongoing development and it is uncertain which ones will progress to Ph3

N.R. Patel (✉)
Cytel Inc., 675 Massachusetts Ave, Cambridge, MA 02139, USA

Center for Biomedical Innovation, Massachusetts Institute of Technology, Cambridge, MA 02139, USA
e-mail: Nitin@cytel.com

S. Ankolekar
Maastricht School of Management, Endepolsdomein 150, Maastricht 6229 EP, The Netherlands
e-mail: ankolekar@msm.nl

Z. Antonijevic (ed.), *Optimization of Pharmaceutical R&D Programs and Portfolios: Design and Investment Strategy*, DOI 10.1007/978-3-319-09075-7_11

trials. Thus budget commitments to drugs that are ready for Ph3 trials have to be made with the recognition of the opportunity cost of reducing these amounts from the budget available to drugs that will enter Ph3 development later in the planning period. The second is the converse problem: if we set aside budget allocations for drugs that we expect to enter Ph3 trials in the future they may fail to progress to Ph3 due to inadequate performance in earlier studies. This frees up funds for deployment but, since budgets are allocated one or two times in a year, there will be delays in deployment of released funds to drugs entering Ph3 development. This can significantly reduce the financial return of the portfolio. To meet the challenge Persinger proposes an approach that considers downstream portfolio level implications of decisions made at the project level along with more frequent portfolio reviews. In this chapter we develop a mathematical formulation that embodies the approach he proposes by modeling the entire stream of allocation decisions made over the planning horizon to optimize financial returns and assess risk at the portfolio level. We also show how this model can be used to re-optimize budget allocations to projects when estimates are updated and assumptions are revised. Using the model for re-optimization provides a structured and consistent process that facilitates rapid and more frequent adjustments to budget allocations over the planning horizon. The models we describe in this chapter build on work described in Patel et al. [1]. This reference also provides a summary of previous work in mathematical modeling of portfolio decision-making in biopharmaceutical companies.

Approach and Mathematical Models

Extending the Traditional Portfolio Approach to Budget allocation

The traditional portfolio approach as outlined in chapter "Challenges of Portfolio Management in Pharmaceutical Development" uses the following inputs to allocate a given budget to projects in the drug development pipeline over a planning horizon (see Sharpe and Keelin [2]). We focus on the budget earmarked for Ph3 development. For each drug in the Ph 3 pipeline (indexed by i) we consider a number of possible design alternatives (indexed by j). For design alternative (j) of each drug (i) the traditional portfolio approach uses the following inputs to determine budget allocation.

1. Budget requirement for Ph 3 development (b_{ij}).
2. Estimated probability of technical and regulatory success (PoSTR_{ij}) in a Ph3 trial.
3. Estimated Expected Net Present Value (ENPV_{ij}) calculated by combining POSTR_{ij} with revenue and cost models.

A ranking algorithm is used to select drugs to develop and to allocate funds to them to maximize ENPV given the available budget. The algorithm computes the return for each drug-design combination as $\text{ENPV}_{ij}/\text{b}_{ij}$ and for each drug chooses the design that gives the highest return, $r_i \equiv \mathrm{Max}_j\ (\mathrm{ENPV}_{ij}/b_{ij})$. It then ranks the drugs from highest to lowest r_i and allocates budget by this rank order until the available budget is allocated.

The above ranking algorithm can be improved by formulating budget allocation optimization as a Knapsack Problem (Martello and Toth [3]) with side constraints. The side constraints are needed to ensure that at most one design can be selected for each drug. Denote the decision variables by Z_{ij}, where $Z_{ij}=1$ if drug i is allocated budget required by design j and $Z_{ij}=0$, otherwise. The Knapsack Problem formulation is given below.

$$\begin{aligned} &\text{Maximize } \sum_i \sum_j \mathrm{ENPV}_{ij} Z_{ij} \\ &\text{subject to:} \\ &\sum_i \sum_j b_{ij} Z_{ij} \le B \\ &\sum_j Z_{ij} \le 1 \quad \text{for} \forall i \end{aligned} \tag{11.1}$$

where B is the total budget available over the planning horizon.

Optimization Model

To address the challenge described in the introduction to this chapter we develop an optimization model that extends the traditional portfolio model to provide, not just an optimum solution at a given time, but an optimum policy that recognizes the downstream consequences in terms of future availability of drugs and the possibility that funds may be freed up by failure of drugs in the development pipeline to progress to Ph3 trials. The extension is based on:

1. *Bayesian priors*

   Prior distributions for the effect size in Ph3 trials are estimated for each drug that is ready at the present time for Ph3 trials, as well as for drugs that are at earlier stages of development in the pipeline that would need to be funded over the planning horizon. These distributions are based on expert opinions or data and simulations from Ph2 or earlier studies. (See, for example, Chuang-Stein [4], O'Hagan et al. [5], Spiegelhalter and Friedman [6].)

   For simplicity we will assume that the prior distributions are two point distributions. This has been found to be adequate in several settings. (See, for example, Cong and Beckman [7] and Patel and Ankolekar [8].) This assumption can be extended to more complex situations where there are multiple levels of success such as when the revenue from launching a drug that has regulatory approval depends on efficacy estimates due to the comparative effectiveness relative to current drugs on the market. See, for example, Patel et al. [9].

   Let Ψi be the prior probability that the effect size of drug i meets the targeted difference from placebo, δ_i, for market launch and Φ_{ij} be the power of design j

for drug i. We assume that design j involves running two concurrent, identical trials that will both need to show significance at level α for regulatory approval.

For the two point prior $\Pr(H_1 \text{ is true}) = \Psi i$ and $\Pr(H_0 \text{ is true}) = 1 - \Psi i$, so that the Ph3 posterior PoSTR for drug i and design j can be calculated as:

$$\text{PoSTR}_{ij} = \Phi_{ij}{}^2 \Psi i + \alpha^2 (1 - \Psi i).$$

We can now compute:

$$\begin{aligned} \text{ENPV}_{ij} = &\{\Phi_{ij}{}^2 \Psi i + \alpha^2 (1 - \Psi i)\}\{\text{NPV}_{ij} \text{ of cash flow if drug } i \text{ is launched}\} \\ &-\{1 - \Phi_{ij}{}^2 \Psi i - \alpha^2 (1 - \Psi i)\}\{\text{NPV}_{ij} \text{ of Ph 3 development cost for drug } i\} \end{aligned} \tag{11.2}$$

This is an important computation as it explicitly recognizes the uncertainty that we have regarding the efficacy of a drug before beginning Ph3 trials. The traditional portfolio approach assumes a value for PoSTR without explicitly modeling the uncertain knowledge of effect size.

2. *A simulation model*

We describe a simulation model to assess technical, regulatory, and commercial risk in the section "Assessing the Risk Associated with the ENPV-Maximizing Portfolio".

We use the Bayesian priors to develop a mathematical programming model that optimizes the drug development program by determining the optimal design options for trials in a portfolio of drugs in the pipeline for Ph 3 development over a planning period. We show how Bayesian decision analysis can be effectively implemented as a Stochastic Integer Programming (SIP) model (Birge [10]). Using SIP, one can optimize decisions that are discrete in nature by modeling budgeting constraints, fixed costs, sequencing and scheduling requirements, and many other aspects that arise in the context of optimizing development of a portfolio of drugs.

The SIP formulation is given below.

Assume, without loss of generality, that the drugs are indexed in the order in which they are scheduled to become available for Ph3 development. So that if t_i is the time when drug i become available for Ph3 development, then $t_1 \leq t_2 \leq t_3 \ldots \leq t_I$. We will assume $t_1 = 0$ and will discount all cash flows to this point in time.

$$\begin{aligned} \text{Let } a_i &= 1 \text{ if Drug } i \text{ becomes available for Ph3 development} \\ &= 0 \text{ if Drug } i \text{ is not available (fails to progress to Ph3 development)} \end{aligned}$$

$$\text{Let } p_i = \Pr(a_i = 1).$$

Here we will assume independent Bernoulli distributions for a_i. (The independence assumption is easily relaxed by modeling the Bernoulli probability of availability of drug i as dependent on the availability of drugs 1, 2, …$i-1$ and replacing p_i by $p_i | a_1, a_2 \ldots a_{i-1} = \Pr(a_i = 1 | a_1 a_2 \ldots a_{i-1})$)

We assume drug 1 is available, so that $p_1 = 1.0$
Decision variables will be denoted as:

$$Z_{ij} = 1 \text{ if Drug } i \text{ trials use design } j, \\ = 0 \text{ otherwise, for } i = 1,2$$

$$Z_{ij|}a_2 a_3 \ldots a_{i-1} = 1 \text{ if Drug } i \text{ uses design } j \text{ when it has an availability history of } a_2 a_3 \ldots a_{i-1} \\ = 0 \text{ otherwise, for } i = 3,4,\ldots I.$$

Data for the SIP model are:

$ENPV_{ij}$: ENPV from drug i if it becomes available for Ph3 trials
b_{ij}: Budget allocation required by Design j for Drug i
B: Total budget available over the planning horizon

The Objective Function to be maximized is given by

$$\sum_j ENPV_{1j} Z_{ij} + p_2 \sum_j ENPV_{2j} Z_{2j} + \sum_{i=3}^{I} p_i \sum_{a_2} \sum_{a_3} \cdots \sum_{a_{i-1}} \sum_j \left\{ \Pi_{m=2}^{i-1} p_m^{a_m} \left(1 - p_m\right)^{1-a_m} \right\} ENPV_{ij} Z_{ij|a_2 a_3 \ldots a_m} \quad (11.3)$$

The first term is the contribution to the portfolio ENPV by Drug 1. The second term is for Drug 2 and is similar to the first term except for the multiplication by p_2 that is needed to compute the expectation over the random variable a_2. The last term is the sum of $(I-2)$ terms reflecting the contributions of the drugs $3 \ldots I$ respectively. The expression in curly brackets is the probability of availability history a_2, $a_3 \ldots a_i - 1$, the expression following the curly brackets is the ENPV contribution of Drug i for that availability history.

The following design constraints ensure that only one of the designs can be selected for any drug under each availability history.

$$\sum_j Z_{ij} \leq 1 \quad i = 1,2$$

$$\sum_j Z_{ij|} a_2 a_3 \ldots a_{i-1} \leq 1 \quad i = 3,4,\ldots,I \quad a_m = 0,1; \quad m = 2,3,\ldots,i-1$$

The budget constraints are

$$\sum_j b_{1j} z_{1j} + a_2 \sum_j b_{2j} z_{2j} + a_3 \sum_j b_{3j} z_{3j|a_2} + a_4 \sum_j b_{4j} z_{4j|} a_2 a_3 + \ldots \\ + a_I \sum_j b_{Ij} z_{Ij|} a_2 a_3 \ldots a_{I-1} \leq B \; a_m = 0,1; m = 2,3,\ldots,I-1$$

Notice that a solution to SIP gives the optimum values of $Z_{ij|} a_2 a_{3,\ldots} a_{i-1}$ for all i. These values specify design choices for each drug under all possible future availability outcomes that will maximize ENPV for the portfolio. The solution, therefore, constitutes *an optimum policy*.

Assessing the Risk Associated with the ENPV-Maximizing Portfolio

The optimization model of the preceding section maximizes ENPV of a portfolio subject to budget and availability constraints. While ENPV is a commonly used criterion in practice for optimizing the return on our portfolio, it does not account for the downside risk of obtaining low and negative returns. The distribution of NPV for a portfolio enables assessment of its technical and regulatory risk. Obtaining this distribution involves convolution of the distributions of NPV_{ij} for $\{(i,j)|Z_{ij}=1\}$. Denoting G as the event that drug i with design j has technical and regulatory success and NG as its complementary event, each NPV_{ij} has a two point distribution F_{ij} with values $NPV_{ij}|G$ and $NPV_{ij}|NG$ with corresponding probabilities of $PoSTR_{ij}$ and $1-PoSTR_{ij}$. The conditional NPV distribution for a given availability scenario given by $(a_2, a_3, \ldots a_I)$ is the convolution of F_{ij} over $\{(i,j)|Z_{ij}=1\}$. The unconditional distribution of NPV is a mixture of these distributions with mixing probabilities given by $\prod_{i=2}^{I} p_i^{a_i}\left(1-p_i\right)^{(1-a_i)}$. A simple way to obtain this distribution is using Monte Carlo simulation.

Risk associated with uncertainty in revenue is often referred to as commercial (or market) risk. The revenue profile over the life cycle of a drug in the market is generally modeled as a function of a few parameters. We can incorporate commercial risk into the Monte Carlo simulation model for technical and regulatory risk described above by treating these parameters as random variables with distributions that are estimated by market studies and expert opinions.

Illustrative Example

We will illustrate our model using an example of a portfolio with seven drugs awaiting Ph3 development. Our emphasis in creating this example has been to keep it as simple as possible while highlighting the key advantages of the general model described in section "Approach and Mathematical Models".

The planning horizon is 3 years. Trials for each drug can start at the beginning of any month within the horizon. Suppose that Drugs 1 and 2 are available for Ph 3 trials at the present time ($t_1=t_2=0$). Drugs 3 through 7 are expected to enter Ph3 development at later times as shown in Table 11.1. While Drugs 1 and 2 are ready to begin Ph3 trials, Drugs 3, 4, and 5 are in Ph2 development and Drugs 6 and 7 are in earlier stages. Target differences in efficacy from placebo and standard deviations of patient responses for each drug in the Ph 3 trials are given in Table 11.1. The prior probabilities that drugs will have the target effect (ψ_i) are assumed to be 0.5 for all the drugs reflecting historical failure rates in Ph3 trials. The Probability of Success to enter the Ph 3 development stage (estimated from past development experience, simulation and meta-analysis studies, and published information) is 0.1 for all the drugs. Drug 6, in which we are very confident, is the exception. Its probability of entering Ph 3 is 0.9.

Table 11.1 Clinical and financial data

Parameters	Drug1	Drug2	Drug3	Drug4	Drug5	Drug6	Drug7
Mean response (δ_i)	0.5	0.4	0.5	0.4	0.4	0.3	0.25
SD of response (σ_i)	2	1.8	2	2	1.5	1.5	1
Trial fixed cost (f_i, \$K)	2,805	15	525	2,125	240	125	500
Per patient cost (c_i, \$K)	11	17	25	24	26	15	14
Fixed setup cost (F_i, \$M)	50	500	400	300	500	300	1,000
Contribution (R_i, \$M/month)	175	85	400	200	45	250	500
Exclusivity period (T_i, months)	108	120	135	180	155	180	145
Month when drug becomes available for Ph3 trials (t_i)	1	1	3	6	13	18	25
Enrollment rate (λ_i, patients/month)	20	30	90	45	60	90	45
Treatment period per patient (tp_i, months)	0.3	1	12	12	24	6	12

Table 11.1 shows the financial parameters for the drugs. We will assume that the time for regulatory approval and setup for manufacturing, sales, preparation for launch, etc. is 6 months for each drug.

Trial Designs

Although multiple arms and more complex designs can be incorporated in our model, we focus on trial designs with fixed sample sizes that have two arms: the drug under investigation and placebo. For simplicity we assume a balanced design, although imbalanced designs with fixed sample sizes could be easily included. We assume a regulatory setting that requires two identical trials to demonstrate significant difference from placebo to obtain approval.

We consider six sample size options for each drug in the portfolio. These correspond to a sample size of zero to reflect the option to not include the drug in the optimum development portfolio (i.e., to out-license or co-develop it with expenses for doing so not having to be met by the Ph3 development budget) and the following five power options $\varphi_{ij}=0.8, 0.85, 0.9, 0.95, 0.99$ for sample size if it is included.

The total sample size n_{ij} for a clinical trial with balanced design ($n_{ij}/2$ patients on each arm) for a continuous efficacy endpoint with a normal distribution is given by

$$n_{ij} = 4\sigma_i^2 \left(Z_{\alpha/2} + Z_{\beta ij}\right)^2 / \delta_i^2$$

where, σ_i is the known common standard deviation of response for drug and matching placebo. Z_γ is the upper γ-quantile of the standard normal distribution with α as the two-sided significance level (typically 0.05) and $(1-\beta_{ij})$ is the power.

Calculation of ENPV

We calculate ENPV on the basis of economic factors determined by the choice of sample sizes and scheduling of the clinical trials in the portfolio.

Cash Flow Model

The cash outflows associated with conduct of a clinical trial in our model consist of a fixed cost f incurred to set up a trial at the start of the trial at time k, and a variable cost component related to n patients enrolled at a rate of λ per month. Assuming a per-patient treatment period of tp months and a per-patient cost c spread uniformly over that period, the total patient cost amounts to $c \times \lambda$ per month during a trial period of $n/\lambda + b$ months. At the end of the clinical trial at time $T_a = k + n/\lambda + \text{tp}$, a fixed setup cost of F is assumed to be incurred, if the trial is successful.

In our setting, cash inflows result from sales of the drug when trials are successful. We will consider net cash inflows that result from the difference between the revenue and the variable costs that are incurred to produce and sell the drug. This contribution, R, from sales of a drug will vary over the life cycle of the drug. A typical time profile of contribution would involve an early growth phase followed by a period it is approximately flat, after which the sales decline as newer more effective drugs enter the market, or as the patent on the drug expires, or as competition introduces similar drugs into the market. We will assume that there is a period, the exclusivity period (T), typically the remaining life of the patent, during which the drug is the sole drug in its target market. We will also assume that after a fixed setup time delay, s, at the conclusion of the trial for regulatory submission work and gearing up for production, distribution, and sales, the contribution jumps to its peak value at time period $T_d = T_a + s$ and remains at that value for a period $t_x = \max[T - T_d, 0]$ until the end of the exclusivity period. Since the bulk of the profits for new drugs come during the exclusivity period, we will assume that the contribution drops to zero when it ends. We will not model growth and decay periods or model the effects of competition on revenue. If a cash flow model that incorporates these and other factors is available it can be used in place of our simple models. While more elaborate models may be necessary in a specific application, they distract from illustration of the key ideas behind our models.

ENPV

We designate the time point to which all cash flows are discounted as time = 0 and assume that it will take 1 month from that time to start our first trials.

Let NPV_{ij} denote the NPV resulting from choosing n_{ij} as the sample size with continuous discounting at monthly rate ρ. NPV_{ij} is a random variable that takes one of two values depending on whether the trials result in regulatory approval or not. Let $\text{NPV}_{ij}|\text{G}$ and $\text{NPV}_{ij}|\text{NG}$, respectively denote the value of the random variable NPV_{ij} given a regulatory approval outcome of "Go" (denoted by G) or "No Go" (denoted by NG).

Extending the notation for cash flows developed earlier by adding subscripts i, j, k as required to reflect dependency on the drug, sample size and trial starting time for Drug i respectively, we have

$$\text{NPV}_{ij}\left|\text{G} = R_i \int_{T_{dijk}}^{T_{dijk}+t_{xijk}} \mathrm{e}^{-\rho t}\mathrm{d}t - F_i \mathrm{e}^{-\rho T_{aijk}} - \left(2c_i \times \lambda_i\right) \int_{k}^{T_{aijk}} \mathrm{e}^{-\rho t}\mathrm{d}t - 2f_i \mathrm{e}^{-\rho k}\right. \tag{11.4}$$

where the integrals involving $\mathrm{e}^{-\rho t}$ over respective ranges of time refer to continuous discounting (Bodie et al. [11]) at a rate of ρ per month of periodic revenue contribution at a rate of R_i per month and trial cost at a rate of $c_i \times \lambda_i$ per month. The continuous discounting of fixed costs f_i and F_i involve multiplier $\mathrm{e}^{-\rho\tau}$ where τ is the time associated with incurring those fixed costs. The last two terms in (11.4) constitute the present value of the trial cost incurred for both Phase 3 trials, irrespective of the approval outcome G and NG. If the trial outcome is NG, these terms will become $\text{NPV}_{ij}|\text{NG}$,

$$\text{NPV}_{ij}\left|\text{NG} = -\left(2c_i \times \lambda_i\right) \int_{k}^{T_{aijk}} \mathrm{e}^{-\rho t}\mathrm{d}t - 2f_i \mathrm{e}^{-\rho k}\right. \tag{11.5}$$

Combining (11.4) and (11.5) we have for the Phase 3 program as a whole,

$$\text{ENPV}_{ij} = \text{PoSTR}_{ij} \times \text{NPV}_{ij}\left|\text{G} + \left(1 - \text{PoSTR}_{ij}\right) \times \text{NPV}_{ij}\right|\text{NG}. \tag{11.6}$$

Return on Investment

Let us begin by computing the maximum ENPV we can achieve for each drug. For Drug i this is $\text{Max}_j\{\text{ENPV}_{ij}\}$. The maximum ENPV that can be achieved for the portfolio is $\sum_{i=1}^{7} \text{Max}_j \left\{\text{ENPV}_{ij}\right\}$. For our example, this is \$11,197M and the corresponding budget required is \$293M. This gives us a ROI (defined as ENPV/Budget used) of 38.3 (=11,197/293).

We can use the SIP model to compute the maximum ENPV that can be achieved for different budget levels over a range below \$293. This shows us how the maximum ENPV for the portfolio varies with the size of the budget (B). For our example we display a plot of this relationship in Fig. 11.1. The slope of the line joining the

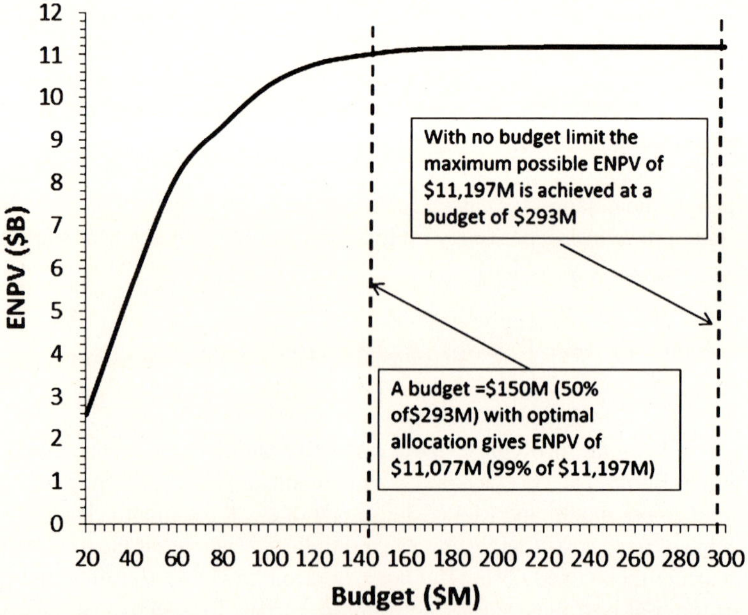

Fig. 11.1 Portfolio ENPV as a function of budget

origin to a point on this curve for a given budget gives us the best ROI we can achieve at that budget.

Note that we need an efficient algorithm to find the best dynamic allocation for a given budget (one point on the curve). This is because with seven drugs and five designs, there are more than 800,000 possible allocations.

Notice that if we reduce the budget by about 50 % to $150M the maximum ENPV goes down by only 1 %. We can increase ROI from 38.3 to 73.8 (=11,077/150) by using an optimal dynamic strategy to allocate Ph3 budget instead of a static strategy that maximizes portfolio ENPV by maximizing ENPV for each drug individually. The ROI is almost doubled!

In fact we can increase ROI even more by reducing the budget further to $100M which results in an ROI of 100 but we will now have a lower maximum ENPV of $10,295M, a reduction of 8 % from the Maximum ENPV for a $293 budget. Figure 11.2 shows the trade-off between ROI and Maximum ENPV in deciding on the level of budget to deploy for Ph3 trials. If one has a target ROI we want to achieve we can determine the budget required to maximize ENPV while meeting the target by using the budget where the target ROI intersects the ROI curve. If, for instance, we have a target ROI of 90 we need a budget of $120 to obtain the maximum possible ENPV of $10,784M.

The best budget allocation strategies for $B = 293$M and $B = 150$M are shown in Tables 11.2 and 11.3 below.

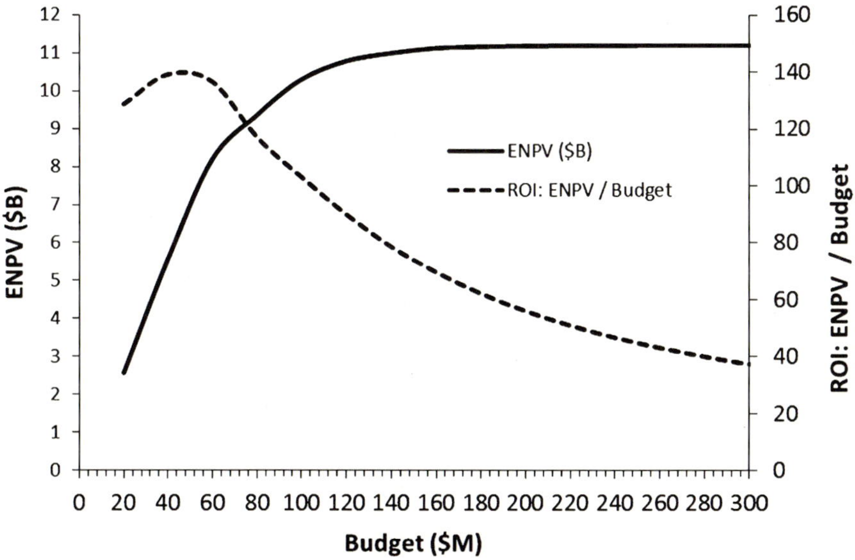

Fig. 11.2 Trade-off between ROI and maximum ENPV

It is interesting to see that with a budget limited to \$150M, the optimal policy is to out-license Drugs 4 and 5 if Drug 3 is available; and also to out-license Drug 5 if Drug 4 becomes available. We have not considered the potential profit from out-licensing in calculating ENPV in Table 11.3. The chance of an out-licensing situation occurring is small here (about 1 %) so it is unlikely to make a significant difference to the optimal dynamic policy. Out-licensing as well as in-licensing opportunities can be readily incorporated into the SIM model as we show in section "What-if-Analysis."

The dynamic adjustment of budget allocations based on the availability history of drugs for Ph3 development can be seen clearly in the case of Drugs 6 and 7 for which the budget allocations that maximize ENPV individually for each drug are \$55M and \$24M which power the trials at 0.99 and 0.95 respectively (Table 11.2). If Drug 3 is available the optimal dynamic strategy scales back power of Drug 6 to 0.95 with a budget allocation that is reduced to \$39M. In that case Drug 7 trials are also scaled back to a power of 0.85 with an allocation of \$17M. If Drug 3 is available but Drug 6 is not, Drug 7 trials are powered at 0.95, so the budget allocated is \$24M, the same as in the static policy of Table 11.1. The SIP considers all such contingencies to determine the dynamic strategy that maximizes ENPV.

To gain further insight into how optimal dynamic budget allocation leads to a major reduction in budget with very little impact on the overall ENPV, let us consider the scenario where all drugs except Drug 6 are available. In this case we see from Table 11.2 that we need a budget of \$238M to ensure that each available drug

Table 11.2 Optimum allocation for budget = $293M

Drug	Contribution to ENPV ($M)	Maximum budget ($M)	Optimum sample size (power)
1	2,628	21	674 (0.90)
2	1,364	28	852 (0.90)
3	839	61	1,176 (0.99)
4	383	66	1,300 (0.95)
5	42	38	732 (0.95)
6	5,292	55	1,838 (0.99)
7	649	24	832 (0.95)

Table 11.3 Optimum allocation for budget = $150M

Drug	Availability scenario	Scenario probability	Contribution to ENPV ($M)	Maximum budget ($M)	Optimum sample size (power)
1	–	1	2,628	21	674 (0.90)
2	–	1	1,364	28	852 (0.90)
3	–	0.1	822	43	832 (0.95)
4	D3 is not available	0.09	311	47	898 (0.85)
	D3 is available	0.01	Out-license/partner		
5	D3 and D4 are not available	0.081	33	31	592 (0.90)
	Otherwise	0.019	Out-license/partner		
6	D3, D4, and D5 are not available	0.6561	3,858	55	1,838 (0.99)
	Otherwise	0.2439	1,428	39	1,300 (0.95)
7	D3 and D6 are available	0.009	52	17	576 (0.85)
	D4, D6 are available, D3 is not	0.0081	43	15	504 (0.80)
	Otherwise	0.0829	538	24	832 (0.95)

has the Ph3 sample size that maximizes ENPV individually for each drug resulting in an ENPV of $23,127M. Dynamic allocation with a maximum budget of $150M as given in Table 11.3, results in an ENPV for this scenario of $18,706M. The difference of $4,421M seems large. However, the chance of this scenario is very small (0.00001), so the reduction in ENPV over all scenarios due to this difference is only $0.04M. On the other hand, if we consider the much more likely scenario where Drugs 1, 2, and 6 are the only ones available, the optimal dynamic policy in Table 11.3 results in the same ENPV of $9,872M for this scenario as the allocation in Table 11.2.

Technical and Regulatory Risk

We have shown that in our example we can increase ROI from 38.2 to 73.8 by using an optimal dynamic strategy to allocate Ph3 budget instead of a static strategy that maximizes portfolio ENPV by maximizing ENPV for each drug individually. In this section we compare these strategies from the point of view of technical and regulatory risk by comparing the NPV distributions using Monte Carlo simulation as outlined in section "Assessing the Risk Associated with the ENPV-Maximizing Portfolio." We ran 50,000 simulations to obtain the probability density functions (pdfs) shown in Fig. 11.3 below. The *x*-axis employs different scales for negative NPV values (scaled in $M) and positive values (scaled in $B) since negative NPV values range in magnitude from 0 to $200M while positive NPV values go up to $80B. A common scale would simply show a single spike near origin.

We can see that for positive NPV values the pdfs for budgets of $150M and $293M are very similar. Although it is not visible in the figure, the right tail for the $293M budget stretches further out than that for the $150M budget. The probability of making a loss (NPV<0) is about the same for both budgets (0.16 and 0.17 for budgets of $293M and $150M respectively). However, from a downside risk perspective, we can see that the probability of a loss exceeding $120M is much smaller for the lower budget. Since the *NPV distribution is multi-modal and far from Normal, it is clear that standard deviation is not a reasonable measure of risk*. The

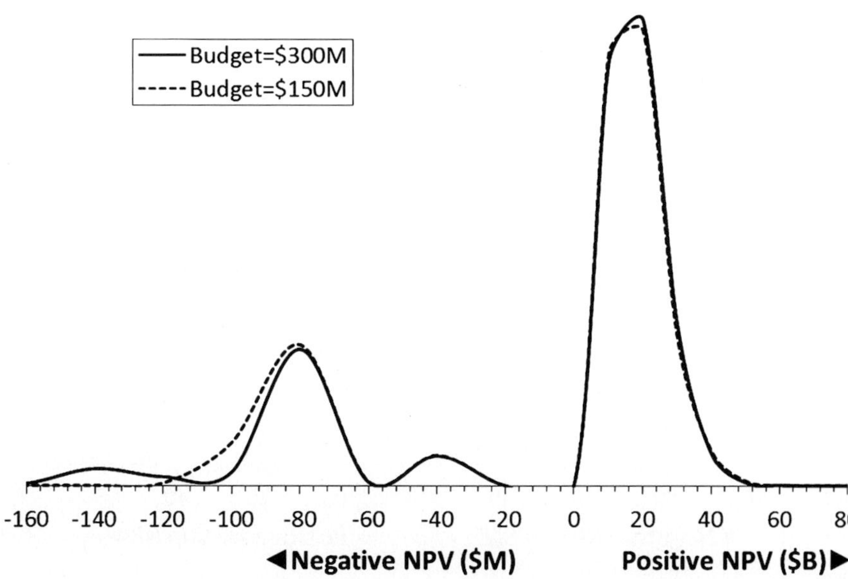

Fig. 11.3 Probability density function for NPV with technical and regulatory risk

Value at Risk (VaR) measure of risk used in portfolio analysis (Bodie et al. [11]) is more appropriate here. VaR is the lower α-percentile of the distribution with a typical value for α being 1 %. Using this measure the VaR for a budget of $293M is $144M compared to $117M for a budget of $150M, i.e., about 23 % higher.

Commercial Risk

As we showed in section "Assessing the Risk Associated with the ENPV-Maximizing Portfolio," in addition to technical and regulatory risk we can also model commercial risk. For our example we do so by treating peak revenue as a random variable instead of a fixed numerical value. Let us suppose that the revenue contribution, R_i given in Table 11.1 is the mean of a Normal random variable truncated at zero. The coefficient of variation (ratio of deviation to mean) of the untruncated Normal distribution is taken to be 1.0, reflecting a fairly large uncertainty in the revenue. In this case 50,000 Monte Carlo simulations of the optimum allocation policy for a budget of $150M shows that the probability of loss has increased to 0.22 from 0.17. Figure 11.4 below compares the probability density functions for NPV with and without commercial risk and Fig. 11.5 shows the corresponding cumulative distribution functions. We notice that the mode of the positive NPV values drops from $20M to $10 when we include commercial risk. Also, with inclusion of commercial risk, the distribution is more skewed to the right. This is because higher revenues become possible when revenues exceed their mean values. For example, the probability of exceeding $28B in NPV has more than doubled from 0.04 to 0.10.

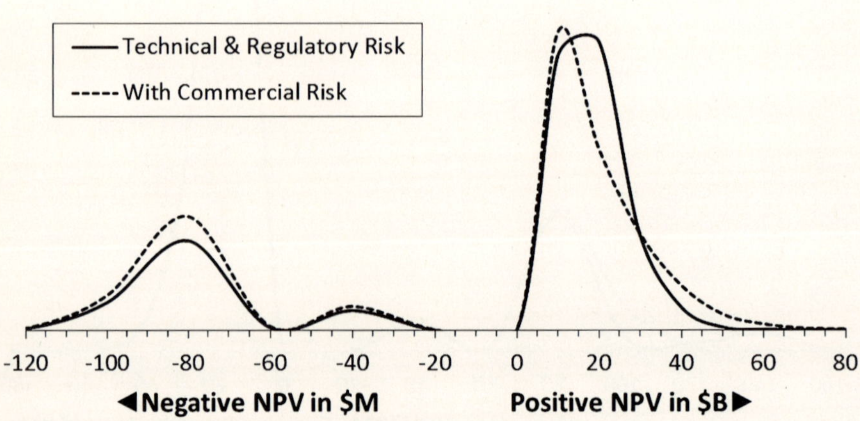

Fig. 11.4 Probability density function for NPV

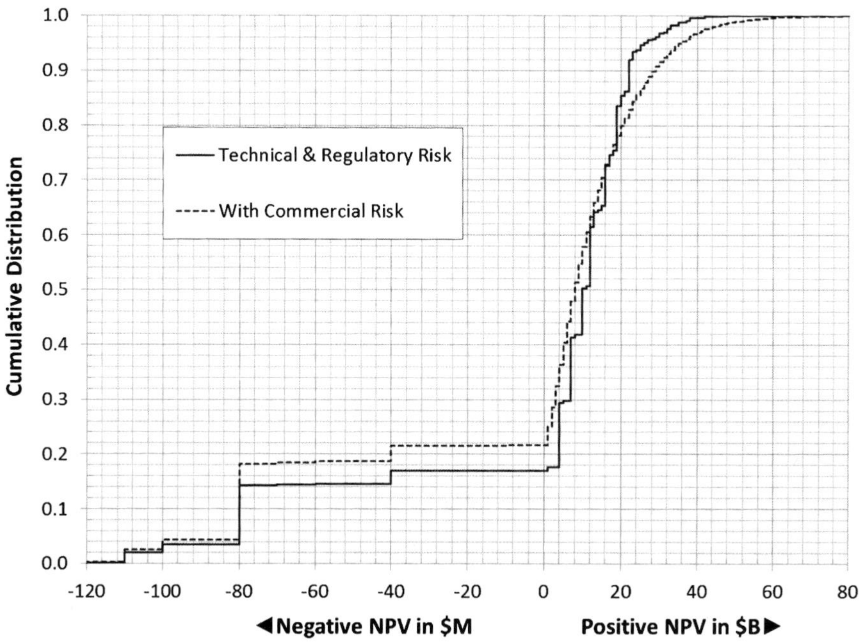

Fig. 11.5 Cumulative distribution function for NPV

What-if-Analysis

A major advantage of the SIP formulation is that it can readily provide answers for a number of important what-if questions. We will illustrate this capability by discussing out-licensing and in-licensing decisions and negotiations in the following sections.

Out-Licensing a Drug

Since Drugs 4 and 5 are to be out licensed in the optimal dynamic policy under certain histories of drug availability when budget is $150M, let us consider out-licensing them outright. From Table 11.3 we see that contributions of Drugs 4 and 5 to the overall maximum ENPV of $11,077M are $311M and $33M respectively. This seems to suggest that we should out-license them only if we get a net ENPV of $344M from the agreement. When we run the SIP without these drugs with a budget of $150M, we obtain a maximum ENPV of $10,747M so the decrease in maximum ENPV is $330M. This is mainly because the new optimal dynamic policy makes up $14M from the budget released to increase the power of Drug 7 trials from 0.8 to 0.95 when Drug 6 is available but Drug 3 is not. This means that if we did succeed in out-licensing them for $344M we would actually earn $11,091M. In fact we could consider out-licensing them for net ENPV less than $344M down to a minimum of $330M.

In-Licensing a Drug

Now consider in-licensing a drug that is under development at another company. We expect the drug to progress to Ph3 testing in month 16 with a probability of 0.2. Our prior estimate of the drug meeting its target efficacy is 0.5. If we develop an ENPV model for the drug as we have done for company drugs and select a set of candidate sample sizes for the in-licensed drug, we can add it as Drug 8 to the SIP model. Let us suppose that the data on Drug 8 is $\lambda_8 = 50/mo$, $tp_8 = 0.25$ *months*, $c_8 = \$23K/patient$, $f_8 = \$1390K$, $F_8 = \$100M$, $R_8 = \$300M/mo$, $T_8 = 165$ *mo*, $\delta_8 = 0.25$, $\sigma_8 = 1.8$ This addition will yield an optimal ENPV for the portfolio of $12,086M an increase of $1,009M over the case when we did not have Drug 8 in the portfolio. The optimum allocation of the budget of $150M is shown below in Table 11.4 with the changes made to accommodate addition to the portfolio without increasing the budget shown in *bold* letters.

Table 11.4 Optimum budget allocation for in-licensing scenario

Drug	Availability scenario	Scenario probability	Contribution to ENPV ($M)	Maximum budget ($M)	Optimum sample size (power)
1	–	1	2,628	21	674 (0.90)
2	–	1	1,364	28	852 (0.90)
3	–	0.1	822	43	832 (0.95)
4	D3 is not available	0.09	311	47	898 (0.85)
	D3 is available	0.01	Out-license/partner		
5	D3 and D4 are not available	0.081	**30**	**27**	**506 (0.85)**
	Otherwise	0.019	Out-license/partner		
8	D3, D4 are not available	0.162	**886**	**34**	**674 (0.95)**
	D4 is available, D3 is not	0.018	**84**	**24**	**486 (0.85)**
	Otherwise	0.02	**103**	**28**	**546 (0.90)**
6	D3, D4, D5, and D8 are not available	0.52488	3,086	55	1,838 (0.99)
	D8 and either D3 or D4 are available	**0.0324**	**161**	**27**	**898 (0.85)**
	Otherwise	0.34272	2,006	39	1,300 (0.95)
7	D3, D6 available, D8 is not	**0.0072**	**42**	**17**	**576 (0.85)**
	D6, D8, and at least one of D3, D4, D5 is available	**0.004878**	**Out-license/partner**		
	D4, D6 available, D3, D8 are not	**0.00648**	**34**	**15**	**504 (0.80)**
	Otherwise	**0.081442**	**529**	**24**	**832 (0.95)**

Notice that the contribution to the maximum ENPV of the additional drug, Drug 8, is \$1073M (=\$886M+\$84M+\$103M). However, the overall increase in ENPV is less than this amount. This is because we include Drug 8 in the optimal policy by pruning back power for trials of Drug 5 trials from 0.9 to 0.85, and of Drug 6 from 0.95 to 0.85 if Drug 8 and either Drug 3 or Drug 4 become available. In addition Drug 7 is to be outsourced if Drugs 6, 8, and at least one of Drugs 3, 4, 5 are available.

Mergers, Acquisitions, and Partnering

Mergers, acquisitions and partnering arrangements are often made in today's environment. These involve evaluation of complex agreements for in-licensing, out-licensing, and cross-licensing of several potential drugs. The SIP model provides a valuable tool for comparing and devising offers and supporting negotiations by extending the approaches described in the previous two sections.

Re-optimizing Budget allocations as Time Progresses and Information is Updated

Now suppose we are at month 7 and Drug 3 has failed to qualify for Phase 3 trials but Drug 4 did make the cut in month 6. Following our strategy we would have started Drug 4 Ph3 trials with the optimal sample size of 898 as shown in Table 11.3. Further, we now have more information on Drugs 6 and 7 from experience at earlier development stages. The Drug 6 development program is delayed by 5 months, so that it will become available is month 23 instead of month 18. Also it is not as likely as we expected to progress to Ph3, the chance of doing so are now estimated to be 0.6 instead of 0.9. Drug 7 on the other hand looks much more promising than it did earlier: its chance of progressing to Ph3 has gone up from 0.1 to 0.5. Let us suppose that, because of having Drug 4 in Phase 3, management decides that the budget can be increased by \$60M, so the budget over the 3 year horizon has increased from \$150M to \$210M. Rerunning the SIP with these changes in input data, we find that the maximum ENPV that can be achieved is \$14,041M which is 27 % higher than the maximum ENPV of \$11,077M that we had expected earlier. The increased budget and availability of Drug 4 for Phase 3 trials along with the increased probability of entering Phase 3 for Drug 7 have caused this increase despite the reduced outlook for Drug 6. (If there was no increase in the budget, the maximum ENPV that can be achieved is \$13,686M). The optimum budget allocation is given in Table 11.5 below.

Using Monte Carlo simulation of the updated optimal policy we find that the technical, regulatory, and commercial risk of making a loss of 0.22 at month 1 has dropped to 0.16 at month 7. The probability of the portfolio NPV exceeding \$20B has increased by 50 % from 0.20 at month 1 to 0.30 at month 7.

Table 11.5 Optimum budget allocation for re-optimization in month 7

Drug	Availability scenario	Scenario probability	Contribution to ENPV ($M)	Maximum budget ($M)	Optimum sample size (power)
1	–	1	2,628	21	674 (0.90)
2	–	1	1,364	28	852 (0.90)
4	–	1	3,457	47	898 (0.85)
5	–	0.1	42	38	732(0.95)
6	D5 is not available	0.54	2,974	55	1,838 (0.99)
	D5 is available	0.06	330	39	1,300 (0.95)
7	–	0.5	3,246	24	832 (0.95)

Summary

We have developed a Stochastic Integer Programming (SIP) model that maximizes the value of a portfolio over a planning horizon by determining the optimal designs of Phase 3 (Ph3) trials for a given budget using Expected Net Present Value (ENPV) as criterion. The SIP model incorporates uncertainty regarding availability of drugs in the pipeline. The SIP provides an *optimal policy* that specifies the optimal design for each drug for every possible scenario of availability of future drugs for Ph3 trials. It optimizes the trade-off between committing budget allocations to drugs available for Ph3 funding at a point in time and preserving budget for drugs in the development pipeline that will need to be funded in the future but whose availability is uncertain. This important trade-off is not handled in a consistent, quantitative way in portfolio budgeting models that are used today.

We have also developed a simulation model to assess the technical, regulatory, and commercial risk of the optimal budget allocation policy.

We have shown how our models can be used for dynamic re-optimization of the portfolio when changes in the internal and external environment occur and as new information becomes available. This capability will enable rapid, frequent and consistent realignment of the strategy to optimize future use of the budget available for reallocation.

We have illustrated our approach by a detailed study of an example that shows how our models can be used to decide on the best budget level to meet a target Return on Investment (ROI) and to evaluate the downside risk of the allocation strategy associated with this budget level. We have shown how these models can be used to answer important what-if questions such as those that arise when in-licensing or out-licensing drug development.

Extensions

Given its flexibility, our models can be extended to handle several important aspects of portfolio optimization. We list some of these below.

1. We can include budget allocations and design choices made in Phase 2 (Ph 2) in the SIP by defining decision variables for Ph 2 similar to those in Ph3 and linking them through constraints to corresponding Ph3 decision variables. This model can be run with a common budget constraint for both Ph2 and Ph3 development or with separate budget constraints for Ph2 and Ph3 to provide an insight into apportioning funds between these phases (see for example Antonijevic et al. [12]).
2. We can perform valuable sensitivity analysis easily by running a series of SIP models. Similarly, modifications of parameters in our models can be used to perform what-if computations to support decisions. We have described such what-if computations for in-licensing and out-licensing decisions in section "What-if-Analysis". Another example would be to vary the impact of market uncertainty for a drug by varying the value of the net revenue parameter, R_i.
3. We have modeled R_i as dependent solely on regulatory approval. However, payer coverage decisions may suggest that R_i will also depend on the magnitude of the effect size, δ_i. This can be incorporated into our framework by modeling R_i as a function of δ_i. Thereafter, we replace R_i by the expectation of $R_i(\delta_i)$ in the first term of (11.4) with the expectation being taken over the conditional posterior distribution of $\delta_i|G$ computed by updating the prior of δ_i. Using this value of NPV_{ij} in (11.6) will reflect the influence of effect size on $ENPV_{ij}$. Thereafter, the SIP model can be used to optimize the portfolio.
4. We can incorporate the impact of safety concerns and adverse side effects on Probability of Technical and Regulatory Success by using a discrete bivariate model for safety and efficacy with an appropriate prior. We can reflect the impact on R_i by modeling it as a function of both δ_i and the safety endpoint (see, for example, Patel et al. [9]). The expressions (11.2) and (11.3) will then involve bivariate calculations. However, these calculations will need to be done only once outside of the SIP computations.
5. If there are other limiting resources besides funds such as scientific, clinical, or project management capacities the SIP formulation can be readily extended to include constraints on these resources.
6. We can also use constraints to take into account strategic considerations such as balancing budget allocation to different therapeutic areas or prioritization (e.g., requiring that a particular drug get funding before another drug is funded for Ph 3 testing).
7. We can reflect the impact of competition if we can model when a competing drug is likely to come to market and the extent to which it would reduce market share for a drug in our development plan. We would modify $ENPV_{ij}$ for the affected drug using this model.
8. If we wish to treat recruitment time for trials as a random variable, we can calculate $ENPV_{ij}$ by averaging over the distribution of the recruitment time. Using these values in our SIP model will provide optimum budget allocations that account for uncertainty in recruitment times.
9. The formulation we have used for SIP may not be computationally adequate for problems with more than 14 drugs. We have made limited experiments using SIP formulations which use "independent scenario formulations" coupled with

"non-anticipatory" constraints because there are powerful algorithms available for computing or approximating optimal solutions for SIPs with this formulation [10]. We have also developed dynamic programming and branch-and-bound formulations that are promising for larger problems. We are studying heuristic algorithms which can provide solutions close to the optimal relatively quickly.

Acknowledgements We are indebted to Kraig F. Schulz of Ernst & Young, LLC for help in constructing a realistic dataset for our example and to Jaydeep Bhattacharyya of Cytel Inc. for assistance in testing our computer programs.

The SIP model for our example has been implemented in the mathematical programming language, AMPL [13]. An Excel front-end, SolverStudio [14], was used to run the model using CPLEX [15] and Gurobi [16] solvers on a PC. The model was also tested on the publicly available Neos server [17] on the Web using the "AMPL on Neos" feature of SolverStudio with the Cbc solver [18]. We sincerely thank the providers of these software tools.

References

1. Patel NR, Ankolekar SA, Antonijevic Z, Rajicic N (2013) A mathematical model for maximizing the value of phase 3 drug development portfolios incorporating budget constraints and risk. Stat Med 32:1763–1777
2. Sharpe P, Keelin T (1998) How SmithKline Beecham makes better resource-allocation decisions. Harv Bus Rev 76(2):45–46
3. Martello S, Toth P (1990) Knapsack problems. Wiley, Chichester
4. Chuang-Stein C (2006) Sample size and the probability of a successful trial. Pharm Stat 5(4):305–309
5. O'Hagan A, Stevens JW, Campbell MJ (2005) Assurance in clinical trial design. Pharm Stat 4:187–201
6. Spiegelhalter DJ, Friedman LS (1986) A predictive approach to selecting the size of a clinical trial based on subjective clinical opinion. Stat Med 5:1–13
7. Chen C, Beckman RA (2009) Optimal cost-effective designs of phase II proof of concept trials and associated go-no-go decisions. J Biopharm Stat 19:424–436
8. Patel NR, Ankolekar S (2007) A Bayesian approach for incorporating economic factors in sample size and design for clinical trials and portfolios of drugs. Stat Med 26:4976–4988
9. Patel N, Bolognese J, Chuang-Stein C, Hewitt D, Gammaitoni A, Pinheiro J (2012) Designing phase 2 trials based on program-level considerations: a case study of neuropathic pain. Drug Inf J 46:439–454
10. Birge JR, Louveaux F (2011) Introduction to stochastic programming. Springer, New York
11. Bodie Z, Kane A, Marcus A (2012) Investments, 10th edn. McGraw-Hill-Irwin, New York City, NY
12. Antonijevic Z, Pinheiro J, Fardipour P, Roger JL (2010) Impact of dose selection strategies used in phase II on the probability of success in phase III. Stat Biopharm Res 2(4):469–486
13. http://ampl.com/products/ampl/ Accessed on May 13, 2014
14. http://solverstudio.org/ Accessed on May 13, 2014
15. http://ampl.com/products/solvers/solvers-we-sell/cplex/ Accessed on May 13, 2014
16. http://ampl.com/products/solvers/solvers-we-sell/gurobi/ Accessed on May 13, 2014
17. http://neos-server.org/neos/ Accessed on May 13, 2014
18. https://projects.coin-or.org/Cbc Accessed on May 13, 2014

Index

Z. Antonijevic (ed.), *Optimization of Pharmaceutical R&D Programs and Portfolios: Design and Investment Strategy*, DOI 10.1007/978-3-319-09075-7

Printed by Printforce, the Netherlands